Seeds of Change

Seeds of Change

Five plants that transformed mankind

Henry Hobhouse

Sidgwick & Jackson
LONDON

A note on conversions

In British pre-decimal coinage
12d = 1s = decimal 5p
20s = £1 = 100p

1 kg = 2.2 lb
1 lb = 0.45 kg

1 billion = 1000 million

Throughout the book it is assumed that
£1 = $1.25
$1 = 80p

First published in Great Britain in 1985
by Sidgwick & Jackson Limited

Copyright © 1985 Henry Hobhouse

Sketch maps and line drawings by Milne Stebbing Illustration

ISBN 0-283-992263-3

Typeset by Rapidset and Design Limited
Printed in Great Britain by
The Garden City Press, Letchworth, Hertfordshire
for Sidgwick & Jackson Limited
1 Tavistock Chambers, Bloomsbury Way
London WC1A 2SG

Contents

Introduction

This is a very personal book, with an original theme. The idea that is central to *Seeds of Change* stems from my impatience with the usual cause-and-effect explanations of human conduct in history, which are generally very unsatisfactory. History is full of the exploits of men and women – their actions are identified as causing change, development, catastrophe. If it pleases people to read and to be told that these things happened as the direct and absolute intervention of men then this is no surprise – for men have always liked to believe in their own influence and direction in the course of history.

These claims, however, sometimes conceal the truth. This book is about an unexpected source of change which has hitherto been obscured because man has been looking too closely at his fellow men. That vital and largely unrecognized factor in the historical process is not man, but plants.

Why did the Mediterranean peoples cease to dominate Europe? What led Europeans subsequently to spread all over the globe in post-Renaissance times? The starting point for the European expansion out of the Mediterranean and the Atlantic continental shelf had nothing to do with, say, religion or the rise of capitalism – but it had a great deal to do with pepper. In the Middle Ages pepper was essential to flavour otherwise insipid vegetables and to mask the taste of rotten meat and stinking fish – when the wrecked Royal Navy ship *Mary Rose* was raised from the seabed in the early 1980s, nearly every sailor who went down with her in 1545 was found to have a little bag of peppercorns in his possession. By the beginning of the sixteenth century Venice had become rich and beautiful from the profits of the pepper trade, which was a Venetian monopoly; however, since about 1470 the Turks had been impeding the overland trade routes east from the Mediterranean, and by 1520 they had nearly brought the Venetian spice trade to a halt. As a result the great Portuguese, Italian and

Spanish explorers all sailed west or south in order to reach the Orient. The Americas were discovered as a by-product in the search for pepper.

All this is known to historians, but its importance in the great European exodus is under-rated. So important – but at the same time so pedestrian – was pepper as a cause, that in the five hundred years since the Turks first created the problem historians have invented all sorts of other reasons for the European voyages which pepper inspired. What relatively new nation, such as America, wants to be offered such prosaic motives for its discovery? What child wants to know that it is merely the outcome of an apparently casual affair?

If pepper is not credited with much influence upon world affairs, nor are the plants in this book. Who has ascribed enough importance to quinine, sugar, tea, cotton or the potato? Other plants, too, have played their part in history, and are still doing so. Tobacco, for instance, an addiction which swept into fashion in some countries, was the essential import which corrected the chronic balance of payments deficit of the American colonies in 1774. Did it therefore finance the Revolution? It is an attractive theory. But if tobacco had not existed, would not something else – timber, grain for the West Indies, dried fish for southern Europe – have taken its place, as indeed happened aftert 1783? In our own time, the transfer of maize from America to Africa has helped to provide a staple food for perhaps 100 million people, and is at least as significant as the Irish potato story. But maize, unlike the potato in Ireland, has not always been grown to the exclusion of all other crops, and whereas the Irish were weaned away from the potato in the fifty years after the famine of 1845–6, the African maize story is still continuing. Having looked, therefore, at these and many more plants and plant transfers, it became clear that it was the five discussed in this book which have played the most important roles in history and, as a result, have most influenced the world in which we live today.

Quinine cured one of the great banes of existence in Europe, Asia and West Africa: malaria. It enabled the white man to open up the tropics and develop great empires. White settlement became possible – only because of quinine – in areas which hitherto had been defended more ably by disease than by any human agency. The second great effect of quinine was to facilitate the transfer of vast numbers of people as cheap labour – probably over 20 million in all. Without quinine, they would have died in their new home. Tamils in Sri Lanka, for instance, and East Indians in Africa, Fiji and the West Indies now form minorities, often troubled and disaffected, sometimes 12,000 miles from the land of their ancestors a century ago. Who attributes their displacement to quinine? Thirdly, the shortage and expense of natural quinine set off the search for synthetics, today diversified into a vast range of industries. Without its acknowledged world leadership in this field Germany would probably have been unable to fight either of the two world wars.

Sugar cane, taken by the white man to the West Indies at the time of the Renaissance and cultivated on plantations by black slaves – the only people who could work in the climate – was the cause of the infamous transatlantic slave trade which made the Caribbean more negro than Africa. All this was for the sake of a product which is wholly superfluous in the diet, a luxury when expensive and a menace when cheap. The brutality of the treatment of the slaves could not possibly be matched by the toothache, the heart pains or the obesity of the consumer. Yet the consumer's misery is undoubted, since after the acknowledged drugs sugar is probably the most damaging of the commonly consumed addictive substances, and is known to be associated with at least one type of cancer.

Tea followed the spices as the major Eastern trade; by 1700 it had become one of the great non-alcoholic drinks and with coffee and cocoa, which had a calmer history, ruled the drawing-room for 250 years. (Before World War I very few soft cold drinks were drunk, due partly to lack of refrigeration, partly to the high cost of pre-motorized transport, and partly to suspicion of liquids which had not been boiled or sterilized.) Tea was obtained at a cost which accelerated the decline of China, a country whose civilization was highly sophisticated when Europe was inhabited by barbarians and America many centuries away from discovery. The exchange of opium for tea over more than a century was a crime which no one even today acknowledges as the man-made catastrophe it was. Tea was more than an incident in the American War of Independence, was instrumental in the development of porcelain in Europe and China, permanently influenced sailing ship design, and by transfer in the nineteenth century developed 'gardens' in India and Sri Lanka which changed the history of the sub-continent.

Cotton, taken to the uplands of the Southern States of the USA, gave declining slavery a new lease of life and provided a cash crop, a political-economic *raison d'être* for Dixie. Seventy-five years later cotton and the slavery issue drove the United States into the Civil War, the War between the States. It was an incompetently managed, savage affair and in terms of loss the greatest conflict between the Napoleonic Wars and World War I – though it was the Homeric struggle that was apparently necessary to forge the nation.

Almost everyone connects the potato with Irish history, in particular with the great famine of 1845–6. But there are other questions that are not always asked: why was Ireland, alone of all western European countries, peculiarly suited to the potato – or was it? Why did the Irish adopt the potato at all? Why did the population rise as steeply as it did – creating ideal conditions for famine? Why did the British espouse free trade as an apparent answer to Ireland's problem? Why, after the mass emigrations, did particular areas of the USA become Irish – what might be called the greening of America? Did the British have any idea of what the Irish tragedy and free trade did to them? The consequences of the potato are still being experienced today in the turbulent inter-relationships between Britain, Ireland and the USA.

If we dismiss these fundamental influences on the course of history we deny the kind of truth which every observer of nature has to admit. The world cannot evolve solely through actions consciously willed by man. Nature can halt our progress and nature can advance it, and man would be foolish to think that his role is any more than that of a propagator of the seeds of change.

Henry Hobhouse
August 1985

Quinine

Quinine
and the White Man's Burden

In 1638, in the Viceregal Palace at Lima in modern-day Peru, the beautiful wife of the Viceroy lay very ill. She had malaria. It was of the intermittent kind, and the recurring crisis of cold-dry, hot-dry, hot-wet was repeating itself apparently hopelessly and to one end. The countess's husband, Don Luis Fernandez de Cabrera Bobadilla y Mendoza, fourth Count of Cinchon, consulted the court physician yet again. Greatly daring, the physician in desperation suggested the use of the northern Andean remedy, quinquina. Bark was procured from Loxa (or Loja), 500 miles away in modern Ecuador, and the countess was cured.[1] In her honour, the genus of tree from which quinine is extracted was named *Cinchona*.[2]

The story is important not only because the countess was the first European on record to have been cured with quinine, but also because, on her return to Europe in the 1640s, she employed quinquina bark to contain and perhaps cure the endemic fevers in the lands surrounding her husband's estate at Cinchon, about 25 miles southeast of Madrid. The Cinchon estates were bordered by the Rivers Tagus, Jarama and Tajuna, and the land, though potentially productive, was often swampy and ill-drained. Given a more vigorous workforce, free from malaria, the swamps could now be drained and replaced by rice paddies.[3] Little did anyone involved know of the episode's significance. Though it was the precursor to great advances in chemotherapy and the whole science of synthetic chemistry, to much tropical settlement and involuntary migration, the event was hardly noticed at the time.

In 1753, in his original great book of botanical record, *Genera Plantarum*, Linnaeus misspelt 'Cinchon'; the mistake was only corrected in 1778 by two Spanish botanists, Ruiz and Pavon. By then, the direct line of the countess's descendants had died out. The castle of Cinchon was destroyed by Napoleon's armies in the Peninsular War. The sleepy town still slumbers, unaware of its place in history, off the main road, on the way to nowhere.

★

In 1981 the World Health Organization officially abolished smallpox; that is, WHO triumphantly claimed that the disease no longer survived. Sceptics viewed this medical rhetoric with some reservation, because twenty years previously the same high hopes had been expressed about malaria. But since the 1960s malaria has made a comeback, and this disabling disease has once again begun to affect all the wetlands of the world.

The word 'malaria' comes from the Italian *mala* (bad) and *aria* (air), and is of nineteenth-century origin; earlier, malaria was known as swamp fever and by various other names.[4] The vector of the disease and the life cycle of the protozoa which the vector transmits to man are now well known, but the true story only became apparent in the present century.

Briefly, the vector is a mosquito; the malarial type is of the family Culicidae, and of the genus *Anopheles*. There are nearly four hundred identifiable species and subspecies, but only sixty are vectors of malaria. The male *Anopheles* is an innocent; he lives on nectar and fruit juices and bothers no one. The female malarial mosquito can only lay her eggs on stagnant water. She must also previously have sucked warm blood, which she must usually do at night. By day, gorged and satisfied, she rests. In sucking human blood, the mosquito infects that person with the blood of the previous human bitten.[5]

Three elements are necessary for the successful continuance of the disease: mosquitoes of the correct type, stagnant water and infected humans. Remove one of these factors, and malaria is 'abolished'. In hilly areas, for example, mosquitoes are still hard at work, but sucking clean, pure blood and therefore having no more than nuisance value. In areas where stagnant water has been drained or covered with a film of kerosene, the mosquito has nowhere to lay her eggs. Elsewhere mosquitoes have been eliminated by insecticides. But all these conditions are only temporary: bring the three together again, and the disease will once again rage worldwide.[6]

Certain humans, notably West African negroes, are immune to malaria because of the peculiar protection afforded by their blood group to the destructive action of the parasite upon their red corpuscles. In most humans, the haemoglobin in the red corpuscles is converted by malaria to melanin; the whole system is affected by fever, and paroxysm follows. Tremors become more and more violent; the face assumes a livid hue; involuntary chattering of the jaw is followed by uncontrollable and increasingly violent shaking of the limbs; fits may follow; the fever, which starts cold-dry, becomes hot-dry; next there is an intermission, like the eye of a hurricane. Then the sweating begins.

The third phase, hot-wet, is succeeded by exhaustion. The patient may then either die, if unattended, or fall into a deep sleep from which he awakes apparently refreshed. The whole cycle lasts from six to twelve hours. Some, but not many, die from the first attack. A few have one attack only, which serves as an inoculation against further infection. Usually, the disease takes the form of a disabling, recurrent fever which may return uncontrollably at any time, set off sometimes by shock, or for no apparent reason. The infection normally remains in the body for life.

The long-term and pathological results of the recurrent fever are the destruction of the red corpuscles and the metamorphosis of haemoglobin into melanin. In layman's language this results in anaemia, melanosis (darkening of the skin), enlargement of the spleen, and permanent damage to the liver – which may ultimately, and often, include fatal cancer of that organ. General debility and an appearance of starvation are typical in severe cases, since the bloodstream is unable to nourish the body.

The only places on earth in which malaria has never been present are the Antarctic and a few islands, such as the Falklands, where an infected human being has never coincided with the right type of mosquito. But the bilge water of ships and the airlines of the world have played a great part in spreading the disease by bringing together mosquitoes and humans. Introduced or imported man-made malaria has been recorded from Archangel in Russia (latitude 64°N) to Cordoba in the Argentine (32°S). Mosquitoes are much less height-sensitive than is generally supposed. Malarial *Anopheles* have been found as low as the Dead Sea (1200 feet below sea level) and as high as 8000 feet in Kenya, or rather higher in Bolivia. In both cases the mountain sites were very close to the Equator.

No one knows how or where malaria originated, but a reasonable speculation would be that the disease will recur anywhere as a result of an infected mosquito, an infected recurrent case (the youngest human carrier alive today is not likely to die until 2050 or later) or an airport in a low-lying location, near stagnant water not sprayed with kerosene. Those who place a strong emphasis on ecological virtue have almost guaranteed the reappearance of malaria by their efforts to discourage persistent insecticides and their campaign against the use of kerosene on stagnant water.[7]

Malaria is dramatically discouraged as soon as the cycle is broken, even temporarily. Drainage; limited range – the mosquito can only fly about two miles; dry, high-lying habitations; drought; nights – the only time the female mosquito flies – cooler than 40°F; and winter are all factors which prevent the spread of the disease. Local factors also help, such as using mosquito nets, leaving very bright lights on in rooms, and sleeping more than 20 feet above ground level. The last has given rise to a number of civilizations in which the survivors have lived in huts on stilts some distance above the stagnant water which is their daytime environment and their source of food: examples are Glastonbury lake village in 200 BC, the Po valley in 100 BC, and some parts of Borneo today.

The destruction of human effectiveness has been vividly described, if insufficiently quantified. The inhabitants of the Tarai region of India, on the edge of Nepal and on the way to Darjeeling, one of the healthiest places on earth, were described thus a century ago: ' . . . miserable, listless, ugly, flat noses, prominent ears, large heads, distended bellies, slender limbs and sallow complexions. The children who survive death from malaria are suf-ferers from malnutrition, sallow skin, enlarged spleen, a huge and dam-aged liver, and sometimes with dropsy.'[8] The condition of the indigenous people today, when malaria has been 'abolished', would be no different but for the counter-measures taken by Europeans in the last century. In the Tarai, and elsewhere, it is certain that malaria will recur.[9]

In Italy there is evidence of more than two thousand years of 'swamp fever'. There have been improvements and regressions since 300 BC. North of Rome in the Campagna, or ancient Latium, which can be regarded as almost the same area, the native Latin tribes were made slaves in their own homelands after their defeat in battle, and worked the land in huge estates called *latifundia*. Drainage was neglected and malaria became endemic. An elegant solution was propounded: make the aristocrats live on their estates. Thus they will cure the fevers or die in the attempt. Cicero and Cato approved. Ostia, once much healthier than Rome, was as bad as anywhere in Italy in the 1880s. Rome itself, malarial in 1750, according to Horace Walpole, was healthy in 1900; the Pontine Marshes, notorious in 1900, were drained by Mussolini in the 1930s. The situation in Venice and her hinterland, which were disease-prone at the end of the eighteenth century, was alleviated by the Austrians in about 1860, regressed in 1870–90, and was transformed again in our own time. The Po valley, a killing ground in Napoleon's 1796 Italian campaign before either side had fired a single shot, improved a century ago, regressed in both world wars, and is always liable to become malarial if preventive measures are neglected.

The invention of the railway not only spread mosquitoes and people, but also allowed the fortunate to escape for the summer from fever-ridden places which they were quite happy to occupy in the winter. Charleston, South Carolina, became a refuge for the surrounding gentry long before the railroad, but afterwards other American cities were avoided during the marsh fever season. New Orleans, Richmond, Philadelphia were all aban-doned during the summer as soon as transport improvements made it pos-sible. Many other cities in a similar situation were deserted by those who could afford to do so. Not unnaturally, since the city fathers were absent in the hills or at the healthy seashore during the malaria season, little was done to improve conditions. It was the Dutch, who often had nowhere else to go, who were the first to drain their cities and make them habitable in high summer. For this reason, Amsterdam, though an ideal site for malaria, was not in fact often afflicted, and the Dutch carried their preventive wisdom with them to their colonies in the East Indies.

All these efforts were made before the life cycle of the mosquito was understood. But book learning is not necessary for survival. No one knows why a new-born calf avoids poisonous plants. No one knows why grass-eating animals ignore grass grown near the excreta of their own species, but relish that grown on the droppings of another, and thus avoid picking up the parasites that affect their particular species. No one knows how from birth a deer can distinguish between a dangerous snake and a harmless one. Likewise, long before man could read and write, the connection between marshes, stagnant water, fever and insects was as deeply embedded in the human consciousness as is a lamb's knowledge that the shepherd's dog is harmless, but that a strange dog is dangerous.

It is not usually until just before a disease is found to be curable, or avoidable, or preventible, that it is identified. Identity, in other words, is very rapidly followed by control. An example is the great nineteenth-century killer, 'consumption'. Was it tuberculosis, or cancer, or psychosomatic? Until the difference between TB and cancer was recognized, it was impossible to analyze 'consumption'; so with ague, swamp miasma, marsh fever or malaria, which was for many years confused with other fevers, notably yellow fever, which is also dependent upon the mosquito, dirt, damp, and a high density of people. The concentration of people is a key to many diseases, and a cycle can be plotted in various parts of Europe during the last two thousand years. If people living in a region predisposed to malaria had a few years of poor harvests, or a rapid rise in population, or bad government, and allowed anti-malarial measures to lapse, they would quickly find themselves in a cycle of deprivation, as it is called today.

All species, including *Homo sapiens*, try to increase the numbers of individuals in their species. This is the aim of reproduction. But nemesis follows excessive success. The vicious circle of fevers–malnutrition–lethargy would immediately follow overpopulation in a potentially malarial region, and it could be said, in a cold-blooded way, that it was as natural a method of population control as famine in an area susceptible to cyclical droughts. A vigorous or enterprising medieval or Renaissance ruler would introduce or reintroduce and enforce good drainage and hygiene and lift a place to a position of economic viability and political importance. Then disaster, in the form of overpopulation, would strike and decline would be repeated. The Nile delta decayed several times in the period of which we have historical knowledge. The same is true of the lower Euphrates and the collapse of the Indus civilization before the Mogul invasion. After the fall of Rome similar cycles occurred in the wetlands between Seville and Cadiz in Spain, in the Rhône delta, at the mouth of the Danube, and, in the case for which there is a great deal more solid evidence, in the coastal belt of Portugal.

The inefficiency, disablement and short life of the survivors rendered whole civilizations victims of their acquisitive neighbours. Feudal peoples were apt to be accentuated victims of this cycle, since the only wealth accruing to a medieval lord was the surplus produced by his land, his animals and his serfs. High numbers, particularly of men, were therefore necessary for defence, for more aggressive military service, and for surplus labour to be devoted to the lord's land for a few days a week or a few months in the year. High numbers encouraged disease, and the cycle would start up again. Nomads often adopted certain patterns of travelling not only in order to find food but to avoid malaria, and the decline of the population density in the world's wetlands is notable after the more severe limitations of serfdom came to an end – once man was not tied to a particular piece of land, he always voted with his feet for health as well as for sufficient food, water and means of shelter. In the census of 1801, the first available in the United Kingdom, the wetlands are notably underpopulated compared with the more healthy districts, and compared with the same parishes in the Domesday Book.

In more modern times, certain military undertakings were rendered far more expensive than they would otherwise have been because malaria aided the defenders. Cartagena in 1741, Walcheren in 1810 and Rangoon in 1824 were all protected by the disease, which cost more in terms of lives than did the defenders. The peaceful occupation of Cyprus by the British in 1878–9 became very expensive in manpower, because the Foreign Office paid no attention to local opinion, which admitted that malaria was endemic. The cost was as high as that of a fully fledged military campaign and even impinged on the army estimates of the following year. Other countries, especially France, laughed at Britain's incompetence. But three years later, in Tunis, the French suffered twenty-five times more casualties from malaria than those inflicted by the human resistance.[10]

In times of peace the malarial death rate was not as great as in wartime, since the population density was less, but it still went on. The disease was endemic in Europe, Asia and probably north of the Equator in Africa. The white man took it all over the world.

In the Americas, the effect was very great. During the early colonization of Virginia malaria killed many more whites than did the Indians, and, to return the compliment, the Indians lost an appreciable proportion of the population of the coastal tribes, perhaps more than half within ten years of white settlement. This recurred throughout the wetlands and may well have been an influence in the enslavement of negroes rather than the malaria-prone Indians. Many of the latter, in the Caribbean, were dead within twenty years of the arrival of the Spanish. The 'pining' of the Indians, noted by every writer, may well have been the result of recurrent and debilitating fever, rather than a fault in the human spirit.

The white man brought malaria to Australia. Though the mosquito can only thrive in certain swampy areas near the coast and cannot breed in the dry interior, it is noteworthy that the aborigines have been adversely affected in those areas which can sustain the fever.

A great deal of information is available from nineteenth-century India, since the British became super-sensitive on the fever issue once it had been identified. When the population was only 150 million, malaria killed a million babies under one year old, another million children between the ages of one and ten, and crippled another 2 million (mostly over ten years old) with recurrent fever. This was in a 'normal' year; in an 'abnormal' year the rate of loss doubled. Taking the lower figure, India was losing 1.3 per cent of its population under the age of ten every year, and that compares with the live births necessary to maintain the population in western Europe today (the net reproductive rate), which is 1.2 per cent. It is difficult to escape the conclusion that malaria was a population control mechanism, and that, but for the advance of science, the natural population of today's India, without either excessive malaria or its control, would be about 100 million. The actual population today is nearly 700 million.

Whites have never seen themselves as spreaders of malaria all over the world: rather the reverse. Natives were always blamed for the disease, even if there was evidence that the white man had brought it with him. Europeans were advised never to build a house within a mile of a native village, and never to allow native children into their own houses. It was thought that native children would be more likely to have the malarial protozoa in their bloodstream than the adults, though no one would know this except by observation. This rule was not followed with negroes in the Southern States, presumably because it had been observed that negroes were immune to malarial fevers.

Apartheid was reinforced by considerations of disease in every area of the world where white men went. It is often alleged that to be separate is wholly cultural, or wholly prejudiced, or wholly sexual in origin. But there was a strong hygienic base to many of the rules. No prudent sea captain would try to enforce apartheid in a strange country with a crew which had been to sea for up to a year, but in those areas where malaria was endemic, and with a crew which had to get back to Europe, discipline was fierce. Sailors stayed on board. In other 'safe' areas, like the South Seas, explorers such as Cook and Bougainville allowed their men ashore. Captain Bligh did not, and everyone knows what happened to him.

The question of where malaria was indigenous in prehistoric times is inadequately documented. There were so many fevers, most of them unidentified. It can be argued that malaria existed everywhere in the tropics, but in that case, why was resistance only developed, to an effective extent, in West Africa? Indigenous malaria in the American tropics would have made impossible the Inca, Aztec or Mayan civilizations. With a death toll as high as is common in malarial regions, there would not have been the surplus of human beings for slavery, blood-letting and sacrifice which evidently existed when the Spanish arrived. Nor would certain Caribbean islands, unexplored by the Spanish, be said by the British to be malaria-free; Barbados, for example, was said to be free of malaria as late as 1660, when it had a white population of over 20,000 proprietors and indentured servants, apart from black slaves who were, of course, immune. Nor would the native Arawaks and Caribs of the West Indies have been decimated by 'fever' after the Spanish arrived, if the Spaniards had not brought malaria with them. This is circumstantial evidence, and there is much more, but it can be assumed with conviction that malaria was indigenous in Europe, Asia and West Africa, and probably not elsewhere.

In that case, why did the natives of Lima know about Peruvian bark, which came from the mountains? Perhaps the court physician who cured the Countess of Cinchon was a post-Renaissance man with an open mind; perhaps he knew that quinquina was used for fevers other than malaria (which was also the European habit for three hundred years); perhaps he knew that quinine could reduce temperature as well as mitigate ordinary influenza, for example. So, if there was no indigenous malaria in Peru, what use did the Indians make of the bark? Other fevers, childbirth, post-natal therapy and abortion are on record, both in Latin America and in other parts of the world after the bark had been spread by Europeans. If these arguments are accepted, then Peruvian bark/Jesuits' bark/quin-quina/quinine is a hinge product in another way. It was a cure which was found thousands of miles away from the source of the disease. This is traditionally held to be odd, since human settlement has always depended upon the cure for an ill being within the range of the humans who suffer that ill. Without that provision, the alternative is grim: drought means starvation; a change of climate involves migration for survival; a disease such as the Black Death presents only death. Until quinine was discovered, malaria was just as socially destructive as any plague, and a continuous, nagging nightmare to rulers of all Mediterranean countries. No one knew that the cure was 5000 miles away from the seat of infection.

★

Dr Juan de Vega, the Countess of Cinchon's physician, returned to Spain in 1648. Something of a scholar, he had published in 1636 a grammar of the language of the Indians around Lima. He brought back a quantity of bark and sold it at Seville, a very badly affected part of Spain, at about an English sovereign for every ounce, or about $75 per ounce in today's money. Not unnaturally, the cure was only available to the very rich. Quinquina became known for a time as the countess's bark, but it would have been better called the doctor's old age pension.

Import of the bark became a new trade. In Latin America, the Jesuits were the protectors and friends of the native Indians. It was the Jesuits who arranged collection of the bark in Peru, Bolivia and Ecuador, and powdered it and sold it for the benefit of the Order. About 1650, quinquina started to be identified with the Society of Jesus, and began to be called Jesuits' bark, thereby becoming a controversial item. Jesuits in Europe were not admired or loved by Protestants, whose prejudices prevented them from believing any good of the drug. One of the great Protestants of history, and the most distinguished English victim of malaria, died prematurely as a result of this anti-Jesuit animosity. A sufferer from chronic malaria, he refused treatment all his life and once called quinine the powder of the devil. His name was Oliver Cromwell.

Louis XIV of France, who thought he could make use of the Society of Jesus in his efforts to subdue Jansenists, Protestants and other dissenters in his authoritarian regime, could only obtain quinine extract from the Society. But he was determined to be free of this monopoly. In 1679 he obtained not only a source of supply of the bark, but also the secret of preparing quinine from an Irish–English doctor, Sir Robert Talbor (or Talbot). Sir Robert was well rewarded, being given a French title, a large pension and 2000 *louis d'or* as a down payment. From that moment onwards France, which at the end of the seventeenth century was considered to be in the forefront of every kind of progress, became the centre of the northern trade.

Many people today would regard Louis XIV as having been overgenerous, for the 'secret' was to grind the cinchona bark to a fine powder in a pestle and mortar and mix that powder with white wine. This crude but effective method of making medicinal herbs available to the body was widely used and, indeed, recommended by most herbalists of the day.

Much of the bark was of little value and often adulterated. Rich people would buy a supply of powder and keep it against a future fever, but by the time they came to use it the powder might have lost its force, and it might not have been any use in the first place since there were five different-coloured barks, and more than seventy species and varieties of trees, a large number of them clinically useless. No one knew which barks worked, or what they contained, or why what they contained appeared to ease the suffering of the fever-stricken. Many must have kept their antidote to malaria for years and then found it wanting. In 1810 a shrewd Yankee skipper sold to the British Army in Walcheren, Flanders, a load of quinquina bark. The British were losing a thousand men a day to fever – more than they were losing to enemy action – but neither the Yankees nor the British knew whether the bark was the right one, or why some men recovered and others did not. The American had sold the British a mixed load, but he was no more guilty than the Indians who had collected the bark, or the Spaniards who had transported it to the coast. [11]

The use of quinine by the English aristocracy increased rapidly after Louis XIV's involvement in the trade, for all things French were much admired by them. Quinine also spread throughout the wealthier classes in Germany, the Netherlands and Switzerland. The canny Jesuits kept control of the trade in Spain, Portugal and Italy.

Until about 1900 no one understood how quinine worked, so there must previously have been an element of magic about the whole business. It is remarkable that vast areas of the world were rendered habitable, vast numbers of people saved from recurrent and debilitating fever, vast numbers of children saved to grow into adulthood, vast changes in population, social and economic land use brought about, and all by a few ounces of a white powder whose chemical action was a complete mystery.

The cinchona trees which produce Peruvian bark do not grow only in Peru, but in a narrow swathe, in cool and equable temperatures, on the slopes and in the valleys of the folds of the Andes, from 10°N to 19°S. They do not naturally grow lower than 2500 feet or higher than 9000 feet above sea level. Below the cinchonas grow palms, bamboos and all forms of tropical vegetation, above them a few stunted alpine shrubs. In the wide belt between there are many varieties of cinchona, some valuable, some medicinally worthless.

The Incas had only been installed in Peru for three hundred years at most when the Spaniards arrived in the area in the 1530s. The centre of Inca power was Cuzco, in the high tableland of the Peruvian Andes. Their greatest emperor, the Great Inca Huayana Capac, had died only a year before the arrival of the first Spaniards, under Pizarro. The Incas had conquered a huge tract of western South America, as far south as Santiago in Chile, and this area the Spaniards took over with relative ease. The Incas were illiterate, but numerate, and probably as fine a race of stone carvers as the world has known. Their staple foods were maize in the lowlands, and potatoes above the grain line. They carefully stored food against future emergencies, keeping a tally with an elaborate system of *quipus*, analogous to the binary system used for computers, expressed in knots. They worshipped their State and were as nasty in their subordination of individual freedom to the needs of the State as any modern European dictator. The fact that with only 183 men Pizarro subdued the Indians at the vital time by treachery was not resented. The Incas used that sort of behaviour themselves, and admired their conquerors.

One of the ironies of quinquina is that most South American Indians believed that all human ills were either hot or cold, dry or wet. Thus a disease would be hot–dry, hot–wet, cold–dry or cold–wet. The remedy should be the reverse of the condition. Thus, if the patient suffered from a hot–dry condition, he or she was given a cold–wet drink. The malarial condition with its three distinct phases baffled them. Given at any point in the cycle, quinquina temporarily increases the body temperature as a precursor to the reduction of the fever. This conflicted with Indian doctrine. So prejudiced were they that as late as 1890 they refused the free issue of quinine, believing that the only cure for fever was cooling drinks.

Another irony of the early history of quinquina is that its exploitation was controlled by the intellectually obscurantist. The original collection of the bark was firmly in the hands of the Indians and Spaniards of the Andes, and the trade within Europe was at first an exclusively Jesuit enterprise. This had two undesirable results. The first was that the dead hand of Spanish lay bureaucracy, with all its corruption and inefficiency, made the collection of the bark more difficult. The second, as already mentioned, was that Protestant suspicion of all things Jesuitical limited distribution of the ground bark – apart from a small quantity to a few liberal Protestant aristocrats – to Catholic Europe only.[12] All the great territorial advances made by (Protestant) English, Dutch and American soldiers and settlers up to the end of the eighteenth century were achieved without the benefit of much quinine. Possibly the only place where it was used regularly before Independence in the future United States of America was New Orleans, which had been both French and Spanish.

Until 1780, the only effective bark exported from South America was shipped from the Peruvian port of Payta. It derived from the area of Loxa and was of one variety only, *Cinchona officinalis*. Peruvian bark was expensive. It was 'hunted', not farmed, and the trees were being destroyed in the process. Yet the Jesuits are said to have insisted that, as a religious duty for every tree felled, another should be planted. Like the Ten Commandments, this rule was honoured more in the breach than in the observance, and by 1795 the German naturalist Humboldt noted that 25,000 trees a year were being killed in the region of Loxa alone. This destruction of natural wealth was as serious in proportion as the destruction of the whale today, and Humboldt was as worried as any modern ecologist.

★

Mankind's attitude to any unknown factor in the physical sciences which lies at the core of human advancement was perhaps best expressed by Francis Bacon in 1585: 'We are not', he wrote, 'to imagine or suppose, but to *discover* what nature does or may be made to do.' This is the first recorded statement of an essentially post-Renaissance sentiment which was to trigger the whole European scientific enlightenment. For malaria, however, this enlightenment was to take several centuries.

Since 1640 Spaniards, and then other Europeans and Americans, had known that there existed what they called a febrifuge, quinquina, the bark of the cinchona tree, but more than that they did not know. Until about 1830, everyone merely imagined or supposed what nature did or might be made to do; a few set out actually to discover. There were two lines of investigation: the first was chemical, to find the active constituents of the bark. Crude methods, incorrect conclusions and wrong turnings delayed the progress of the enlightened party, while the reactionaries were able to point out the shaky nature of the whole enterprise. In France in the 1740s it was held, justifiably, that if only half the patients recovered when given quinquina bark their cure could as well be ascribed to magic, or chance, or prayer. No one in the enlightened party had yet discovered which bark was effective.

The first breakthrough was the correlation between fine fibres which cover the transverse fractures in the bark and the effectiveness of the febrifuge in the bark itself. This random stroking of bark gave way to an intense and international analytical effort. In the fifty years between 1779 and 1829 over three hundred published monographs attest to the efforts of men working with primitive methods and often faulty reagents. The chemists involved were French, English, German, Scottish, Russian, Swedish and Dutch. The final puzzle was solved by Louis Pasteur in 1852. It is noteworthy that none of the scientists involved was born in Spain, in whose colonies and ex-colonies all the bark then grew. In the discipline of natural science, Spaniards have been singularly lacking in curiosity.

The answer to the question was aided by the discovery that the bark contains four alkaloids, analogous to strychnine or morphine, in varying proportions: quinine, cinchonine, quinidine and cinchonidine, the last three only slightly less powerful febrifuge agents than quinine itself. For all practical purposes, all four alkaloids were known as quinine.

Once the tests for the four alkaloids could be established, the relative value of the various barks could be properly assessed. Previously an infusion of bark either proved effective, or left the patient to die or to survive on his own. (In the Far East, malaria-ridden coolies were given infusions of bark from a load before the purchaser would agree a price. If, however, rich Europeans were in fever, the haggling was less vigorously pursued.)

There was a brisk market in bark in Paris, Amsterdam and London from about 1780 onwards, and the retail price was about £1 per pound, falling from about 1820 as a result of increased supplies from South America. Demand rose from about 1840 and prices climbed again to £1 per pound. This was equivalent to about £100 per pound in today's money. Since nearly 2lb of bark (of about 4 per cent febrifuge content) was needed to treat a case of malaria, and the product of nearly a pound needed to be taken each week to prevent the fever recurring, it is quite clear that, even after the valuable barks were identified, the method of preparation formalized and the bisulphates of quinine available in any first-class chemist, only the more affluent could afford to protect themselves.

At the other end of the chain, between 1820 and 1850 the 'hunters' in the forests of the Andes were encouraged to find more and more bark of the most valuable species. The bark was cut in the most wasteful way, stripped in such a manner that the tree inevitably died. Trees passed the point of renewal at that time in history when the Spanish colonies, after considerable exertion, became independent during the period 1810–30.

The great Spanish empire south of the isthmus of Panama was never visited by any Spanish emperor and by very few grandees, except for those who went out to govern. There were never very many Spaniards permanently settled in South America, unlike the numbers in Mexico, and the continent was relatively empty, the Spaniards seeking their fortunes in modern-day Peru or Bolivia but returning to Spain to retire and die. Very few pure Spaniards were born in South America in the first three hundred years of the empire, though Spanish seed was sown throughout the continent from northern California to southern Chile. The centre of gravity of the Spanish effort was among the silver mines of Peru and Bolivia, the great mines of Potosi in modern Bolivia having produced between 1550 and 1650, it is said, more than half the world's supply of new silver.

There is no evidence that malaria existed in mainland South America before the white man arrived, but the disease was endemic throughout Europe, and Spain was particularly badly affected. The sea ports and marshes of that country were deserted during the fever season, flight being the only prophylactic. Cadiz was often half empty after the fleets sailed for South America in May, and it is probable that the same wise policy would have been pursued in Peru if there had been any evidence of native and endemic malaria. But in 1638 the Countess of Cinchon had been at sea level at Lima during the fever season, and her disease was only cured by quinquina. So it is probable that the Spaniards brought the disease with them, and the cycle built up into a dangerous syndrome only after the first hundred years of Spanish occupation.

The Andean republics – Columbia, Ecuador, Peru, Chile and Bolivia – threw off Spanish rule by 1820, but by the time they had gained a sort of semi-stable independence, which was not really until the late 1830s, malaria was endemic in low-lying areas in all five countries. However, the cure was now known and widely disseminated, if expensive. The export of bark to Europe was a thriving trade, reaching a figure of at least one million lb in the 1840s, and nearly that figure from Bolivia alone in 1860. The tonnage reaching the United States at the same time is not known, but it must have been high, because the shortage of quinine was one of the great hardships in the Southern States during the blockade imposed by the North in the Civil War.

By 1850 the British had decided that a secure, controlled imperial supply of cinchona was necessary, and that the trees should be grown on a plantation basis, rather than merely 'hunted' in some foreign forest and then traded on an insecure and speculative exchange. The British Army in India alone needed an annual supply of at least 750 tons of bark, or 1.5 million lb, more than was exported and available in the 1850s. To prolong the lives of the 2 million adults dying of malaria in India every year would require ten times as much: 15 million lb. To render economically active the 25 million survivors of malaria who were crippled by recurrent fever would require at least ten times as much again. And then there was Africa, with a fever rate of up to 60 per cent in some regions. There was a world shortage of Peruvian bark and every sign that the forests would soon be exhausted. The British establishment groaned into activity, its intelligence operating at various levels, like some great primeval beast with a brain dispersed throughout its limbs (today this is called 'consensus').

At this date the British in India had moved far from their original trading impulse and become rulers of most of what is now India, Pakistan and Bangladesh. The East India Company, which had lost its monopoly of trade between India and the outside world in 1834, had illogically become a private company governing a sub-continent.[13] The whole anomaly was to be brought to an end after the Indian Mutiny of 1857, and for more than half a century India would become a mere raw material producer and a market for British manufacturing industry. The East India Company had already been forced by British manufacturers in the eighteenth century to squeeze to death the traditional Indian cotton cottage industry, in order to favour the cotton mills in Britain, and as a result the population of Indian manufacturing towns fell by three-quarters.[14]

The displaced workers were driven into the countryside, and a parallel development, quite unconnected, had the effect of turning them into serfs. Beginning in Bengal in the eighteenth century, the Company had farmed out the only inland tax it raised, the land tax, in such a way that the collectors became a hereditary class of 'landlords'. They were converted, in effect, into permanent owners, not only of the right to collect taxes, but also of the land itself and of the tenants, who, loaded with the need to pay out up to half their crop to the tax collector, sank deeper and deeper into debt. Struggling to make enough to meet their basic needs, the peasants neglected communal works such as roads, canals and river control, and not unexpectedly malaria increased dramatically. The substitution of an overcrowded and landless peasantry for a pre-industrial craft-based society made a lot of money for the East India Company in the short run, but it would bring the Company to an end and pile up trouble for both India and Britain for the next century. These mistakes were those not of wicked men, but of thoughtless and pragmatic rulers who could only see life in the very short term. They were supported and influenced by the commercial interest, whose perspective was of even shorter focal length, often unable to look beyond the next harvest.

Having, as a by-product of their land policy, increased the incidence of malaria to a point where it threatened the effectiveness of the whites in service in India, the East India Company looked for a cheap febrifuge, since they were paying out about £100,000 a year on imported cinchona bark which covered only the more valuable citizens of all colours. The demand was expressed to London, where the rulers of India saw no reason why cinchona plants should not be brought from the Andes and naturalized in the south Indian hills, which were of roughly the same height and latitude as the plant's native home.

The second element in the British establishment which had arrived at the same conclusion – the need for a daily dose of 'Peruvian bark extract' for millions of tropical subjects of the British Empire – was the staff of Kew Gardens. Kew Palace was a one-time royal nursery for the numerous children of George III and had been surrounded by an ornamental garden laid out by George's mother, Princess Augusta, and her landlord, Lord Capel. The Gardens were always more important than the Palace, which was a plain, unadorned, seventeenth-century brick building. In about 1840 Kew Gardens became the powerhouse of British botany, including what are probably the finest botanical museum, herbarium and library in the world.

The scientists involved with Kew included the great early entrepreneur, Sir Joseph Banks, who was an untrained, managerial botanist in an era when there was little systematic botany of any kind. The very rich son of a rich father, Banks introduced botanical lectures at Oxford while still an undergraduate in the 1760s. He explored Newfoundland and Labrador, bringing back hundreds of unknown botanical specimens, and accom-

panied Captain Cook on his first voyage to the Pacific, fitting out a ship at
his own expense and hiring draughtsmen, painters and a distinguished
botanist, Dr Solander, to accompany the expedition. Banks' effort transformed the expedition from a mere observance of the transit of Venus into
one of the most valuable explorations of the natural world to that date. He
was unable to accompany Captain Cook to his death in the Sandwich
Islands (modern Hawaii), but, as consolation, explored Iceland and the
Western Hebrides, one of the first Englishmen to do so.[15]

Inevitably Banks became President of the Royal Society, the source,
repository and organic heart of the natural sciences in eighteenth-century
Britain. He was at the time much admired by those in his own intellectual
camp, but, as J.H. Maiden wrote in 1909 in *Sir Joseph Banks*, 'he was
inclined to deprecate the mathematical and physical sections of the Royal
Society, and he exercised his authority somewhat arbitrarily.' This is too
polite by far – Banks was an interfering, opinionated, bossy man, surrounded by all the favourites, toadies and enemies that such people attract.

Among his brainwaves, Banks had suggested the collection, growing
and trial of all the species of cinchona growing in the Andes. This was not
to be done for a generation after his death in 1820, and by the time the
English and Dutch naturalists had got around to a systematic exploration
of the home of 'Peruvian bark' the position in South America had become
difficult, not to say dangerous, for any botanist.

In the thirty years after the wars of independence – from about 1820 to
1850 – the five Andean republics had enjoyed no fewer than forty-three
revolts, uprisings and revolutions, quite apart from those against Spain
which made them independent. Life was cheap. Different factions, often
consisting of the last *caudillo* (leader) or the next, or disaffected generals,
commanded armies which varied in size and which controlled or disputed
various different areas. The western South American climate was affected
by both latitude and altitude, ranging from the tropics to the Antarctic,
from sea level to 20,000 feet. Communications were naturally difficult and
suffered from human interference. The Andean region is larger than the
United States but long and thin – about 5000 miles long and from 400 to
1000 miles wide. Always sparsely populated, the region had probably less
than 1 per cent of the population density of England in 1850. No one
would have accurate figures, but they were small enough for a man to disappear voluntarily for many years, and no one would ever know whether
he was still alive, or had been murdered, or had succumbed to the many
wild animals, snakes or fevers of this great empty land. The native Indian
population, outnumbered by half-castes, had been much reduced by
Spanish colonialism and was concentrated in mining areas. There was a

shortage of food, only partially relieved by imports along the bad roads from Argentina.[16] Brazil was virtually cut off by the swamps and jungles of the Upper Amazon. Almost the only sure way of travelling was by sea, or on land by foot, horse or pack mule. There were few roads in the Andes themselves, and they certainly formed no network. After gold and silver, the most valuable commodity by weight was quinquina bark.

The mountain slopes and gullies were a paradise for the botanist because the steepness of the tortured, folded mountains made for a very rapid change in vegetation. Hundreds of species and thousands of varieties are indigenous to the Andes, but in the 1850s only the Peruvian bark was of economic significance. It was an ambitious amateur, Clements Markham, who persuaded the India Office and the Director of Kew Gardens that to seek out quinquina plants and reproduce them in India would be an imperial enterprise of great value.[17] With characteristic economy the authorities funded the expedition at minimum cost. The whole sequence of plant transfer from Bolivia to southern India probably cost less than £10,000 (about £500,000/$600,000 today); for this sum, the British got a bargain.[18]

Markham, born in 1830, was the son of a canon of Windsor and the grandson of a Yorkshire land-owning baronet. A typical example of early Victorian poor-but-well-connected youth, he joined the Royal Navy at fourteen, became a midshipman at sixteen and a lieutenant at twenty-one. Before he was of age he had been to the Arctic on a fruitless expedition to find the explorer Franklin. At the tender age of twenty-two he retired from the Royal Navy to become a full-time (but amateur) explorer and geographer, which the English have produced in hundreds, most of them failed – usually as a result of early and sudden death – forgotten and unrecorded. Markham travelled by himself in the eastern Andes in 1852–4 and wrote three books based on his experiences. Just after the great shake-up which occurred in India as a result of the Mutiny in 1857, Markham successfully persuaded the India Office in London to let him organize an expedition to South America to bring back alive a number of small cinchona plants to be naturalized in a suitable soil and climate in India. This he did between 1859 and 1862. About half-a-dozen other Englishmen were involved, but they were modest and silent and we do not know of their contribution. Markham is kindly but patronizing about their efforts, and he gives credit to 'Dr Spruce, Mr Pritchett, Mr Cross, Mr Writ, and Mr Ledger', whom he calls, with jovial insincerity, 'my fellow-workers'. A parallel Dutch effort to transfer cinchona plants from the Andes to Java was, in the end, successful (see p. 23) but it appears that many of the wrong varieties, low in febrifuge, had originally been obtained.

Markham's goal was to provide a prophylactic dose 'at a cost of less than a farthing a day' for anyone in India exposed to malaria. This was a revolutionary aim, since during the period when the bark was hunted by Amerindians in the forests of the Andes and cured in London, Paris or Amsterdam, the cost to an East Indian of a preventive dose was nearer 1s a day – forty-eight times Markham's hoped-for price. It was also several times more than the cost of a coolie's daily food, with the result that it was more expensive to protect a workman than to feed him. Distribution was difficult, for the coolies would not spend their own money on quinine: 'They would ràther risk disablement and death than give up their little pleasures.'[19] It is not easy to relate a farthing and a shilling in the India of 1860 to modern realities, but it would be fair, against an unskilled American wage of today, to call them the equal of about 50c and $25, or in Britain 40p and £20. Many weekly wage earners in Detroit or Dagenham today would resent having to spend $25 or £20 a day on a prophylactic. The unemployed would consider it an outrage. In India in 1860 there were many unemployed, many economically inactive and many subsistence peasants. A farthing a day was a noble aim for a Victorian benefactor of mankind; it was also a good slogan, which a politician could appreciate.

After many difficulties and adventures Markham and his allies succeeded, with or without the help of Kew Gardens. Some of the plants went via Kew and were multiplied in the alien greenhouses by the sweet Thames. Some of the roots, plants and seeds were also sent to the Botanical Gardens in Calcutta, where multiplication was pursued. Some, romantically, were taken out by Markham himself, who wrote: 'He who would desire to receive the most pleasant impression of India, on a first arrival, must follow in the wake of Vasco da Gama, and disembark on the coast of Malabar, the garden of the peninsula. . . . Late in the evening, we embarked in a canoe on the Beypur river.' The next day Markham met his Scottish host, McIvor, superintendent of the government gardens at Ootacamund; he had already selected a site for the naturalization of the cinchona plants. Markham, forever damning others with faint praise, called him 'a good practical gardener'.

The Nilgiri Hills, which had been chosen from four possible areas as being suitable for cinchona cultivation, were covered with native vegetation including rhododendrons, berberis, caulteria, lilies, lycopodia, ilex, cinnamon, viburnum, acanthus and jasmine, into which abundance of plant life the English had introduced vegetables and fruit from northern Europe, together with oranges, lemons, limes, bananas, nutmegs, loquats and plantains, as well as tobacco, sugar cane and the European nettle, which grew with fierce abandon and stung with sub-tropical savagery. Given a choice of height above sea level almost anything would grow, from a plant which needed temperatures of over 80°F all its life to one which needed mild frost half the year. The area selected for cinchona was in the temperature belt 45° – 70°F, as in the Andes.

The transfer of the plants from the Andes to Kew, to Calcutta and to the Nilgiri Hills was much assisted, and indeed only made possible at all, by the invention of the Wardian case in 1830. This was the brainchild of one Nathaniel Bagshaw Ward, a physician from London who was a great amateur botanist and Fellow of the Royal and Linnean Societies. Ward placed the chrysalis of a moth in a sealed glass jar to observe its emergence into mothhood; he also noted that a plant (some say two plants) developed, within the sealed environment, from seed into sapling. He had discovered the principle of the unique specialized environment, in which the air plus moisture enclosed in the sealed case produces vapour and carbon dioxide during the day, and oxygen, dew and even frost (depending upon induced temperature) by night. No foreign diseases, fungi or viruses could attack the plant within the case. Air and moisture were sufficient for growth when continually recycled in the sealed environment, and the nutrients in the soil, if enough were provided in the first place, would allow the plant to grow. The rate of change in the plant could be controlled by the external temperature to which the case was subject. Before the days of the steamship and air travel it might take six months to reach Kew from the furthest place on earth, and Dr Ward's invention, which we now call a terrarium, proved essential. One plant lived for over a year on a long, interrupted journey from New Zealand to Kew. The folding Wardian case, made of wood and glass, played a key part in the collection efforts of Kew, in the commercial transfer of plants and in the dissemination of species all over the world throughout the nineteenth century.

Meanwhile, in the Nilgiri Hills, the 'good practical gardener' McIvor developed his cinchona plantations rapidly and efficiently, experimenting with growing trees in full sunshine and in various degrees of shade, pruning and manuring, crossing, hybridizing and grafting. But his great contribution to the domestication of the Peruvian bark was to develop two systems of husbandry which did not involve the destruction of the new plantations and allowed an annual crop which was reasonably constant in quantity. This simultaneously produced a great increase in quality.

McIvor began by coppicing one-sixth of the trees every six years, which resulted in an appreciably higher total production of alkaloids. This technique then gave way to what became known as 'mossing', which involved removing bark from longtitudinal strips of the lower trunk, then surrounding the whole circumference of the trunk with indigenous moss, thus allowing the strips to renew themselves. 'Mossed' bark produced an alkaloid level of about 7 per cent; coppicing, to produce 'quills', produced about 6 per cent; and 'natural bark' about 4 per cent. All values would be dependent upon variety, season, sunshine figures, soil temperature and so forth, as with any other crop anywhere upon earth. But in the chosen sites in the Nilgiri Hills the mean temperature throughout the year is about 58°F, with a maximum/minimum of 70° and 45°. The hills lie about 11°N of the Equator, so that the daylight figures are fairly constant throughout the year, and there is neither a recognizable winter nor, at the chosen altitudes, frost.

Both government and commercial plantations were developed in the Nilgiri Hills, and McIvor went on to select and test varieties and to exchange information, seeds, cuttings and plants with the botanical gardens of Calcutta, Singapore and Kew. He also corresponded in the same way with the Dutch in Java.

By 1880 the cinchona industry was mature. There was a peak of 'hunting' production of about 20 million lb of bark from the Andean republics at that date, after which they could not compete with the Eastern plantations. In India, most production was of a low-level mixture of alkaloids, called totaquinine today and quinetum by the contemporary English and Dutch in the East. This low-level mixture, subsidized to cost only half-a-farthing a dose, was available at any post office in Bengal, the most affected Indian province, by 1880. The amount of totaquinine produced is difficult to assess, but was probably enough to safeguard 10 million people with a dose a day.

There was a curious lack of co-operation among the many British authorities in England, Calcutta, south India, Singapore and Ceylon. There was no co-ordination between government departments, let alone between government and commercial growers, chemists, processors and merchants.

Commercial production of pure quinine was discouraged in India, and proved commercially and ecologically unsuccessful in Ceylon. The British planters in Ceylon had developed a successful coffee industry which by 1870 was almost wiped out by the joint efforts of a native rodent, the Golunda rat, and an imported fungus, *Henileia tastatrix*, which thrived to such an extent in the mountains of Ceylon that it finished what plants the rats had left behind. The planters, desperate to use their land, climate and labour, invested heavily in cinchona bushes, without paying sufficient attention to the biochemistry. Eight to ten years later they found that they could not compete with Java which, through a combination of geographical position and altitude above sea level, produced nearly double the active alkaloid per acre that the planters in Ceylon could. Twice ruined, first by coffee pests and then by the growing habits of cinchona, the survivors turned to tea, which was, and remains, a story of much greater commercial achievement.

In India itself, the government absorbed a high proportion of the product of both private and state plantations. The aim was to protect not only every white man exposed to risk, but also every working coolie on any kind of job which involved continuous labour in malarial regions. This in fact did not cover a very large percentage of the population. The economically inactive were excluded; so were the self-employed, unless they were wise enough to spend their own earnings on the prophylactic febrifuge. Most of them found it impossible.

Those who were protected by this new development were the contract coolies, their wives and families, in high-risk areas. They daily dose was issued, and ingested, every morning in a kind of parade on many tea estates, irrigation works or military cantonments. The other 'natives' who lived thereabouts were not protected. Anyone with the money could, of course, buy quinine at a post office in any village of any size, but most Indians lived at subsistence level and never had money. For the self-employed the charge of a farthing per day represented too high a proportion of their income.

Cinchona was being grown successfully in India, the drug was of proven worth by 1880, and the British Empire had more land suitable for cinchona production than all the non-British tropical land in the world. Yet despite these advantages the free trade instincts of the British establishment led the powers-that-be to leave every development outside India in the hands of private enterprise.

The Dutch, who had begun to transfer cinchona plants to Java at the same time as Markham was active, aimed their production at the European market. Under the energetic and hypercommercial leadership of their botanical director, Dr De Vrij, a quininologist who had started out later than the British, they achieved a higher yield of the purer, more acceptable alkaloid, quinine itself. This was sold all over the world after processing in London or Amsterdam, while the Indian production remained largely in India.

Logic dictated that the Dutch reinforce their natural advantage by the creation of perhaps the only purely Dutch-directed cartel in modern history. With a smooth, wise, pawky and apparently inoffensive air worthy of the contemporaries of Rembrandt, the latter-day Dutch traders formed a combine called Kina. Operating out of Amsterdam, this cartel ruined any non-combine trader who sought to undercut the monopoly. As a result, British production was used almost exclusively in the country where it was grown, and plantations were established in Malaya, Burma, Ceylon, Mauritius, East Africa, Gambia, the Gold Coast and the West Indies, as well as in India.

This situation was arrived at partly by choice, since in each of these countries malaria was a local problem of formidable proportions, and partly because the British and Dutch had always found it easy to co-operate, once the seventeenth-century rivalry had been settled in England's favour.[20] Kina's monopoly certainly did not hurt the ordinary British Empire-builder at all. Supplies of quinine sulphate were available almost anywhere in western Europe at any pharmacist, and no traveller who was anybody moved without a supply. The absence of enlightenment was limited to those without Indian or colonial experience: when, for example, as mentioned earlier, Cyprus, which initially was the joint concern of the Foreign Office and the War Office, was occupied by British troops after its annexation from Turkey in 1879, no quinine was issued to the troops. As a result over 10 per cent of those who disembarked died, and over 50 per cent were permanently disabled. For some years afterwards, quinine sulphate was still only available to people in the Levant by mail from western Europe, Athens or Constantinople.[21]

During this period before World War I, American supplies came in the form of either raw bark from the Andean republics, or refined quinine from Europe. There is no evidence that anyone in the United States found the trade important enough or lucrative enough to engage the attention of any great capitalist. Quinine, like tea, was a product of peripheral trading importance which the Americans were prepared to leave in the hands of foreigners. It was not until Japan posed a threat to the major traded supplies of cinchona that the American government showed any concern for supplies of quinine. The Japanese threat was not to develop until the 1930s, and long before this, in the 1850s, the Germans and British had both begun looking for a synthetic form of the drug.

The discovery of synthetic quinine was assisted by the search for synthetic dyestuffs. Synthetics are achieved by rearranging the architecture of base chemicals so that the molecular structure of the chemical fits the end for which the researcher is seeking. Sometimes the researcher looks for one thing and finds another, and thus it was with the original synthetic dyestuff.

In 1834 aniline, a derivative of coal tar, was shown by the German chemist Runge to produce a brilliant blue colour when precipitated with bleaching powder, but his experiment was not developed commercially. In 1856 eighteen-year-old William Henry Perkin, a student at the Royal College of Chemistry under the great A.W. Hofmann, rigged up a rough and rather dangerous laboratory at home; he was not in fact looking for the dyestuff which was to make him rich, but for a synthetic quinine. Having found the dyestuff, he called it Mauveline – from which the colour mauve entered the vocabulary – and set up a factory to exploit his new product.

Paul Ehrlich (1854–1917), another German, was an intellectual dynamo who systematically examined thousands of potential dyestuffs for their possible use as tracers in the human body. Because they are not fast when washed, natural dyes are unsuitable for all but luxury clothes, and useless as tracers in the bloodstream since the colour is absorbed by the body. Ehrlich became obsessed by the effects of the new, fast, synthetic dyestuffs being produced by the early organic chemists – not only by their direct effects on colouring and therefore on their course and reaction within the body, but also by the side-effects, which were of even greater importance, as it transpired, than the direct ones. Over many years he used a vast number of dyes, some of them developed for this sole purpose, to trace the pathway of various organisms in the human body. The malarial protozoa was identified by Laveron in 1881. Ten years later, Ehrlich used methylene blue to trace malaria in the bloodstream of a German sailor with tertiary malaria, and the sailor was cured. This accident was the first recorded case of a cure by a quinine substitute. It could not be repeated, because until avian malaria was discovered there was no safe way of testing new products; after about 1910, however, quinine substitutes could be tested on canaries and budgerigars.[22]

It is not just the discovery of the first quinine substitute that made Ehrlich such an important figure in chemistry, and indeed in history, but also the far-reaching side-effects of that discovery. He produced a cure for sleeping sickness in 1907, and 606, or Salvarsan, the first synthetic cure for syphilis, in 1910; he improved his technique of serum preparation; he did some valuable work in the possibilities of coal tar-derived chemotherapy in cancer limitation; and he produced a successful method of testing the efficiency of anti-toxins in laboratory conditions, without the use of humans or animals. It was the search for synthetic quinine that led to all these other discoveries and experiments, and as a result Ehrlich left German coal tar-derived chemistry in a position of absolute supremacy. This situation was to help the warlords in World War I and, had it not been for Hitler's paranoid and damaging anti-semitism, would have made Ehrlich's (mostly Jewish) pupils and successors the architects of German victory in World War II. In explosives, fertilizers, pharmaceuticals and synthetic substitutes of all kinds the German chemical industry was able to survive defeat in World War I, poor government and inflation in the 1920s, even the slump, largely because of the technological lead derived from Ehrlich and his pupils. Even today, two generations after his death, Ehrlich's beneficent legacy inspires the synthetic chemical industries of both Germanies.[23]

The American and British chemical industries were less well developed in the 1920s, and depended heavily upon German innovation. This was to change when the Nazis came to power in 1933. Within a year, the scientific and technological community outside Germany began to be as enriched by refugees as it was outraged by the circumstances which drove them to leave their homeland.

German genius for synthetic chemistry was known before 1914, but its triumphs in textiles, rubber and oil only became public outside the industry as a result of shortages of these materials induced by the Allied blockade in two world wars; there was much talk of *ersatz* products or synthetic substitutes. But the German synthetic industry went back much earlier. In 1870 the new imperial government represented the youngest nation in Europe. The nationalism which made this new unified Germany possible was political and idealistic, based on language and sentiment, and the parallel in trade and industry was to produce everything at home if at all possible. Heavy tariffs kept out foreign foodstuffs. German manufacturers were encouraged and bullied into buying domestic products, and they became only too willing to do so. German self-sufficiency was already an important and not always beneficial factor in world trade by 1900. Without their base of forty years of self-sufficiency the Germans would not have been able to contemplate World War I, let alone fight half Europe for more than four years. The 1914–18 war was waged with less than 10 per cent of peacetime seaborne trade creeping past the British blockade. It was an astonishing effort on the part of German industry.

One of the great shortages was of quinine in East Africa. An attempt at reinforcement, a Zeppelin flight from Bulgaria carrying significantly more than 2 tons of quinine, failed. The German forces in East Africa remained short of quinine, apart from the small quantity they were able to steal from the British. All through the war Ehrlich's followers had been working hard at the puzzle of rearranging the structure of coal tar chemicals, but it was not until 1926 that the first synthetic quinine, pamaquin, was proved.[24]

The latest drugs, and the clever compounds of anti–malarial preventives and suppressants and sulphacures, were all developed in a great rush at the end of the 1940s and during the very early 1950s. There has been no completely new development in drug design since that period. None of the modern drugs, with the exception of some of the curative compounds and some of the clever injections which prevent malarial infection for up to three months, in any way replaces quinine. Unlike the synthetics, quinine is not known to have produced or encouraged the development of resistant strains of the malarial parasite, but it is not perfect, for its long–term side-effect is to encourage blackwater fever in tropical countries.[25]

Synthetic drugs, which are often bred or cloned or developed forms of other drugs, seem to encourage the same process in the pest or disease they have been designed to kill or cure. This has happened with weedkillers, antibiotics, insecticides and fungicides, all of which have tended to encourage the development of mutant forms on which their usually lethal action has no effect. In the story of malaria, real quinine has never seemed to behave in this way.

The latest thinking on anti-malarial therapy brings us full circle back to the seventeenth century. Negroes suffer from a type of anaemia, known as the sickle-cell condition, which prevents them having malaria although they may act as carriers.[26]

In 1684 a Portuguese writer in Brazil recommended a cure for syphilis: 'Buy a virgin black girl off a ship, and lie with her for a month, and the cure will be effected.' A piece of cruel, self-indulgent nonsense, the modern reader will no doubt say. But he or she would be wrong. Between 1920 and 1950, when the first effective antibiotics became available, the cure for tertiary syphilis was, surprisingly enough, induced malaria. If a man had bought a negro girl in 1684, and she had been a virgin (and therefore free from venereal disease), he would have acquired malaria from her which would have cured his syphilis. But there is never aught for nought in this world. The girl would in return have been infected with syphilis, and unless he had some quinine at hand, the man might have died of malaria.[27]

The advice of that Portuguese writer was good as far as it went. He did not mention either malaria or quinine. Only now do we know why the cure might have worked. And the great question remained in 1684: how do you cure syphilis in a negro who can't get malaria?

In his justly acclaimed and much admired *History of the World*, J.M. Roberts makes the point: 'The change in world history after 1500 is quite without precedent. Never before had one culture spread over the whole globe . . . by the end of the eighteenth century . . . European nations . . . laid claim to more than half the world's land surface. . . . Only the interior of Africa, protected still by disease and climate, seemed impenetrable.' Dr Roberts does not mention malaria, nor is the word in his index. It is not to denigrate his achievement that this point is made, but to emphasize the necessity for this book in general and this chapter in particular.

Quinine and the synthetic substitutes are such valuable hinge products that it is very difficult to exaggerate the importance of the influence upon history of malaria and its control. The ifs of history are unfashionable and considered an academic waste of time: we are what we are, people say today, and we are what the past has made us, and there is no denying the present, or the past which made the present; we have enough trouble in living through the present, and reaching the future in one piece, without bothering about the past. But the present is the child of the past, which is the father of the future, as well as of the present. Consider what would have happened to world history, first without malaria, and then without quinine.

This is not unreasonable. It is not like asking people to consider life without any mineral fuels, which would have the effect of upsetting all the known laws of physical chemistry and make a topsy-turvy world which could only exist on another planet. Being without malaria is equivalent to being without a disease which kills a woman who conceives for the fourth time or a man who marries more than twice. We do not have these hypothetical killers, but they would both form suitable population control devices, which is the apparent purpose of a great many diseases in both the animal and vegetable kingdoms. (Diseases control population growth and ensure survival of the fittest. This is 'cruel': but if malaria is cured, and the food supply does not increase, then the control will be famine, which is more cruel. Drought also kills mosquitoes, and as a result may increase human survivors, who will die of starvation.) Nor is the idea of no malaria a silly one. There was no syphilis in Europe until 1492. There was no TB, measles or influenza in much of Africa, and none in the Americas or Australia, until the white man came. There was no AIDS until recently. So it is legitimate to assume the possibility of a world without malaria.

A world without malaria would have had a very different past and present. There is a whole highly productive, historically vital zone where malaria has been endemic and other fevers – yellow, lassa or blackwater – are not present. In this wide zone, particularly around the Mediterranean and in the Middle East, the absence of malaria would have changed the world. First, there would have been recurring crises of overpopulation, leading to famines – even more than there actually were – and making starvation, rather than disease, the control. The Arabs would not have stopped at the borders of the dry uplands of Africa, the borders of the Sahel. The Vikings would not only have colonized more than the North Atlantic islands, but also, perhaps, gone much further south than Sicily, and even further east than Kiev. The world would have fallen to those with the very greatest combination of necessary qualities, regardless of the imperatives of 'swamp fever'. The wetlands of the world would have been more heavily populated, on a much more stable basis, and the cooler uplands less important than they have proved. West Africa would be more densely populated by all races, and not only the one which adapted over the centuries to resist malaria – the true negro.

Now imagine malaria without the existence of quinine or its synthetic substitutes. The whole history of the world since about 1650 would have been different.

It is known that the absence of quinine in the first 150 years of white settlement in the Caribbean and the Americas led to a very high death rate among the Amerindians who had never known malaria, and the poor white indentured servants who had known it in Europe though in circumstances less favourable to the mosquito. The lassitude caused by malaria affected all whites in the Southern States in more recent times, and the sudden absence of quinine in the period after the institution of the effective Northern blockade of 1862–5 led to considerable fever in the Confederate Army and contributed to Northern success. Had quinine been available, that bloody marginal battle, Gettysburg, might have had a different outcome. No import figures are available for the blockade years, but it is known that in 1864 the cost of quinine sulphate in terms of gold in the Confederate States was ten times the prewar rate, and this is some measure of what life without quinine must have been like for the majority.

A moment's thought about what has happened with the availability of quinine demonstrates the essential influence of the Jesuits' bark. Because malaria was not contained until the last quarter of the nineteenth century, the negroes of the world had been the favoured labour force throughout the tropics in the previous four centuries. It would be going too far to suggest that the absence of a cheap febrifuge resulted in the choice of the immune negro as the slave race above all others. But it is clear that, as soon as cheap totaquinine was available in the Caribbean and Latin America, East Indians and Chinese were imported as indentured labourers.

Dr Roberts writes of 'the interior of Africa, protected still by disease and climate. . . '. Indeed, before the availability of quinine at a low enough price – that is, not until the 1880s – the race for West Africa was not a possibility: it was as if the course of history had been delayed by the unavailability of quinine. It was not just a problem of the dark interior: within as little as a mile of the high-water mark in a dozen ports of West Africa malaria was still endemic in the 1940s. In East Africa the story was in some ways different. Not only was malaria less common than in West Africa, but the indigenous peoples are not resistant, so that the Arabs, who settled in the eighth century, probably did not take malaria with them into the interior. Certainly, it was not endemic when the whites arrived, and in the inland areas of modern Kenya, Uganda and Tanzania the disease did not become a problem until there was a marked increase in population.

The evidence is only anecdotal, but there does not seem to have been any serious incidence of malaria in East Africa until the white man arrived in force, after the seventeenth century. From before that date only the vaguest of evidence is available, but what there is suggests an absence of swamp fever. The non-immunity of the non-negro blacks of East Africa would seem to prove the point in a longer timescale.

It is in India that quinine has, in the last hundred years, had the most serious effects – both good and bad. Large areas of the sub-continental wetlands became habitable and relatively healthy for whites and Indians alike. Assam was exploited for tea, the Indus valley for rice, and much of southern India for a multitude of crops. Large numbers of male Indians became wage-earners for the first time, leaving the women and children to practise subsistence farming on the tiny plots of land which were all that were available to the peasants. But though quinine made workable millions of acres which had previously been lethal to occupy, the benefit accrued to the whites and not to the indigenous population.

All over India and Ceylon, it became possible to import coolies on indentured contracts and keep them alive with quinine, tied to the plantation for the term of their contract. Huge numbers were transported by dealers within India itself; and in Ceylon the native Sinhalese were supplemented by the more willing Tamils from south India, which has resulted in a permanent problem of racial tension. Nearly half the population of modern Sri Lanka is descended from indentured labourers brought in because quinine made it possible.[28]

The trade – for trade it was – in indentured labour arose because India became a surplus area, overcrowded in many regions as the result of drainage works, the improved food supply, and the increased birth rate made possible by the availability of quinine. So Indian workers were exported to the East and to South Africa, Mauritius, Malaya, Fiji and the Caribbean (notably British Guiana) to replace or reinforce the natives who did not possess the Indian work ethic, or who refused to work for cash when they could enjoy a perfectly happy life without paid work or money to spend.[29]

Whole new industries – sugar in the Indian Ocean and Fiji, tea in Ceylon, rubber and tin in Malaya, and bananas in the Caribbean – and the agricultural development of East Africa were all made possible by the use of quinine. This vast population transfer, less inhumane than slavery, less noted by social historians, has nevertheless had as much effect as the African slave trade. Perhaps the most famous result of this diaspora was the development not of an industry, nor of an ethnic problem, but of an individual: Mahatma Gandhi was permanently affected by his experiences as an Indian immigrant in South Africa.

The Chinese were also transported all over the world once their lives could be protected by quinine. Today Malaysia, Singapore, Hawaii, the East and West Indies and the South Pacific all provide homes for overseas Chinese whose ancestors were transported at so much a head by some latter-day speculator. European settlement in Indo-China, Siam and Burma would have been impossible on the scale on which it occurred in the late nineteenth century without the availability of the daily dose of febrifuge.

The Belgian Congo, French Equatorial Africa, German East and West Africa and the Dutch East Indies were able to be exploited in an organized, white-supervised plantation economy only because of the use of cinchona bark.[30]

The Americans in Panama acted with far greater prudence than did the British in Cyprus, and not very much more recently. In the early 1890s, Panama was just as it had been two hundred years before, riddled with fever, crossed by whites only at speed and heavily dosed with quinine. The railroad across the isthmus had been built after gold was discovered in California, and completed in 1855; more than half those employed on its construction died of 'fever'. Until the Americans took over from the French the building of the Canal, people continued to die, but because the early symptoms of malaria were very similar to those of yellow fever it was impossible to treat yellow fever cases until the disease was so far advanced that it was fatal. The Americans drained the swamps and sprayed the lakes with kerosene – necessary to combat both diseases – gave everyone a daily dose of quinine and were then able to set about curing the yellow fever. It was a triumph of preventive medicine, but without quinine no doctor would have been able to start work, and the Canal Zone, in common with most of the tropics, would have remained a white man's grave.

People are very equivocal about synthetics. The word has a downmarket sound – it is a substitute for the real thing. In fact, some synthetics are much better than the natural product – rubber is a good example, for its synthetic substitute, neoprene, is far better at resisting oil and other corrupting liquids than any natural mix.

Synthetic quinine made World War II possible for the Allies. Without it, there would have been no chance of protecting an additional 25 million men and women in the forces who were brought temporarily to some strange place where malaria was endemic. Without synthetic quinine, Japan would have won the war in the Pacific, and the war in the Mediterranean would have turned out very differently.

Real quinine was the first natural product to be used in sustained chemotherapy; it was also the first for which a substitute was sought in synthetic chemistry. In some ways the natural chemotherapy may be preferable to the synthetic alternative. Quinine is cheaper than the synthetic, and pleasanter in use, with fewer side-effects. For a hundred years we have marvelled at the potential of synthetic chemistry, which has given the world not only the systemic – blood-borne – drugs, which cure sleeping sickness, malaria and most of the ordinary infections, but also fertilizers, textiles, plastics and a hundred and one products familiar in every home.

Yet quinine remains the first generally acceptable and entirely natural systemic drug. Modern man forgets, in his creation of the synthetics, what Bacon said in 1585: 'We are not to imagine or suppose, but to discover what nature does, or may be made to do.'[31]

It is now believed that some Amazon plants contain chemically effective natural agents which scientists have never examined. Yet the Amazon basin is being systematically destroyed, ironically a fate first made possible by the availability of quinine. Have naturally occurring chemotherapeutic agents already been destroyed which could cure cancer, which could be more effective than synthetics in helping depressive states, which could prevent deformed foetuses, or multiple sclerosis, or disablement at birth? Are there many other plants containing cures for the ills of civilization? Should we not find out before we destroy their environment for ever?[32]

Notes

1   This romantic true story formed the basis of an inaccurate and very sentimental novel by a remarkable Frenchwoman. The book, published in French in 1817 and in Spanish in 1827, was one of more than eighty by Mme de Genlis (1746–1830), who was one of the first writer/educators to propound the value of learning by doing rather than by rote. She survived the Revolution, the Restoration and the loss of her husband, lover and most of her property. Her more readable works are probably the *Mémoires*, published in ten volumes in 1825 when she was seventy-nine. They are scandalous, racy, opinionated, feminist and historically inaccurate, but a good read.

2   *Cinchona* = botanical name of the genus.

Quinquina = Amerindian name for the bark, literally 'bark of barks'. Also the word used in Romance languages.

Quinine = English name for the extract. Also the most important of the four alkaloids. For most purposes, all four alkaloids can be called 'quinine'.

Other names, such as Peruvian bark, Spanish bark, Jesuits' bark and Indian bark, will also be found.

3   Rice (*Oryza sativa*) is a cereal, an annual, like wheat but with much higher temperature requirements. A native of India, rice spread eastwards to China, southwards to the East Indies and westwards to Persia. The Arabs brought it to the Mediterranean in the seventh century AD. Rice has a larger genetic variation than any other cereal; some varieties like dry land, but these show no advantage over wheat in subtropical climates. The only rice to grow to advantage in the deltas of the Danube, Nile, Po etc., and all over Italy, the Balkans and Spain, are the wetland varieties. They require floodable arable fields into which rice plants are puddled, with water often 4 inches deep; the water is then drained off gradually. It is the control of the floodwater, or irrigation, which mades rice culture possible and destroys the mosquitoes' breeding ground.

4   The word 'malaria' was compounded by Macculloch and first used in the UK in medical literature in 1827. The earliest connection between swamps, marshes, undrained land and lack of health goes back at least to 1000 BC in Persia and Babylon; the disease must have been endemic in the ancient world. In the optimistic 1960s, when WHO believed that malaria could be 'abolished', it forecast that twenty years later only 4 million people would be badly affected by the disease. The actual figure was 400 million, about 8 per cent of the world's population, which makes malaria quite the most serious health problem (or population control mechanism) today, and throughout recorded history.

5 The necessity to drink blood makes the breeding mosquito a true vampire, but less photogenic than the occupants of our TV screens late at night. The blood is only coincidentally infected with the protozoa of malaria. As far as is known, no animal other than the owl monkey (*Aotus trivirgatus*) can be infected with human malaria; the existence of this connection was only discovered in the 1960s. There is also avian malaria, to which a number of bird species are subject. Otherwise, the mosquito can only be a vector of the disease if it has previously sucked the blood of an infected human.

6 Twentieth-century epidemics, with recorded results:

	No. of cases	*No. of deaths*	*Date*	*Reason*
USSR	10,000,000	60,000	1923–6	Medical breakdown
Ceylon	3,000,000	82,000	1934–5	Excessive rainfall
Brazil	100,000	14,000	1938	Mosquito *Anopheles gamiae* caused epidemic; arrived by air from Africa
Egypt	160,000	12,000	1943–4	Control measures lapsed in delta
Ethiopia	3,000,000	150,000	1958	Excessive rainfall
Southern Asia	Resurgence of disease since 1977. Since 1980 there have never been fewer than 10 million cases in the Indian sub-continent, with more than 100,000 foreshortened lives a year			

7 Oil products and Paris Green (arsenite of copper) were both in use about 1900, notably in Central America – Cuba and Panama in particular. In the end, paraffin and kerosene were found to be the most economical inhibitor of the mosquito's life cycle. Until the advent of the insecticides, pyrethrum and DDT, kerosene was a standard preventive measure; it is far cheaper to spray from aircraft and was still in use in the Canadian Arctic in the 1980s. The disadvantage is that it kills fish by deoxygenating the water.

8 Charles Creighton, MA, MD; lecture quoted in *Encyclopedia Britannica*, ninth edition, 1883.

9 Prosperous, sophisticated city-dwellers should not regard malaria as merely a third world problem. After World War I there were 7000 new cases in France, 500 in the United Kingdom, and more than 1000 in the USA. These were all cases of a disease acquired by civilians who had no contact with returning soldiers except via the mosquito. In our own times, the Vietnam War produced more than 20,000 cases of 'introduced malaria', as it is called, inside the continental United States between 1966 and 1976.

10 No army possessed any services of any consequence aimed at prophylactic medicine or sanitation until long after the mid-nineteenth century. Deaths from preventible, avoidable, curable, unnecessary disease outnumbered immediate deaths from enemy action at least until after World War I. Malaria was one of the

prime killers, even after the bark became available. Cost may have been an important factor, or ease of availability, or identification of the effective bark, or just
plain obscurantist generals.

11   Cinchona has two other biazarre characteristics as a biochemical. First, the
amount of active alkaloid in the bark is a function of altitude. At sea level, the effective alkaloids may be almost absent. At the limit of cultivation, at or near the frost
line, the alkaloids are at their highest proportion of the bark's contents. This variation was not fully understood until the 1890s.

Secondly, the solubility of the powdered bark may prove adequate in the test
tube, *in vitro*, yet inadequate in the malaria patient, *in vivo*. Sulphate of quinine,
long used as a standard remedy, is unabsorbed by a small percentage of people; no
one knows why. Formerly, they either recovered spontaneously or died. The
newer medicines were hydrochloride (made with hydrochloric acid rather than
sulphuric) and the acid hydrochloride, the best absorbed and therefore most valuable of all the salts of quinine. It was not generally available until after 1900.

12   Quinine was officially called 'Jesuits' bark' as late as 1808, when an Act of Parliament was passed limiting its export. This was the year of British involvement in
Portugal, and two years before the invasion of Walcheren. The former did not produce many cases of malaria, the latter perhaps 30,000.

13   The British East India Company, chartered by Elizabeth I in 1600, had gradually become the ruler of half of what is now India, Pakistan and Bangladesh, almost
in a 'fit of absent-mindedness', as Macaulay put it. The other half was ruled by
native princes. The Portuguese, French and Dutch had been excluded from any
real territorial possession. The commercial arm of the Company was mostly a
highly profitable, if poorly and corruptly managed, monopoly in tea, opium, cotton etc. The monopoly trading powers were brought to an end in 1813 for everything except tea, and for tea itself in 1834, and the East India Company wound up
its non-governmental activities, remaining until after the Mutiny in 1857 a joint
stock company governing a sub-continent. (The Hudson's Bay Company, in a
similar position in Canada, gave up its governmental role and kept the commerce
for a hundred years. This was the more profitable, as well as the more logical,
course.) The last employee of the East India Company, a veteran of the Mutiny,
did not die until 1932.

14   The process of restricting cheaper pre-industrial cotton, linen and wool fabrics from abroad, including Ireland and India, started in about 1720 and gradually
became more onerous. Imports of manufactured goods were not allowed free
entry until British manufacturers were in a near-dominant position after the beginning of the Industrial Revolution. Between 1800 and 1850 most manufacturers
agreed to 'free trade' because they had no real rivals. See the chapters on Cotton
and the Potato.

15   Banks brought back from Iceland samples of the extremely luxurious down of
the eider duck, collected then, as it still is today, by local farmers from the nests of
eider ducks, which return year after year to the same place. So many quilts were at
one time called 'eiderdown' in England that it would have required a hundred
times the world's population of eider ducks to supply the contents, if they had all
been genuine.

16   Throughout the Spanish colonial period, which started in 1520 on a settled
basis, the annual treasure fleet returned, usually to Cartagena in modern Colombia, with products of the Mediterranean which the homesick settlers could not
grow on the land – wheat, wine and olive oil. This trade continued to Cuba as late

as the 1890s. Most of the region now draws these essentials from the USA, not Spain. Wine is, of course, produced in Chile and Argentina, but was not produced commercially before they were free of Spanish rule.

17 Kew performed three essential functions in the nineteenth century: a collection of living plants in the Gardens allowed people to see the exotic growing; a collection of dead material permitted the study of plants which could not be grown; and the continual intellectual analysis of new vegetable material brought Linnaeus' work up to date with each discovery, and allotted a position to each new plant brought in by insatiable botanists.

18 Sometimes the economy was overdone. The six 'workers' who helped Markham were not well treated. Their total pensions amounted to less than £2500, included in the £10,000. (Markham, *Peruvian Bark*, John Murray, 1880.)

19 From a letter written by Lady Dufferin, the Vicereine, in 1888.

20 Anglo-Dutch co-operation appears to be more natural than Anglo-French or Anglo-German enterprise. Joint Anglo-Dutch undertakings include, besides quinine, shipping, banking and the great Shell and Unilever companies.

21 Quinine could not be bought retail except in large towns and cities. In Egypt in 1895, for example, quinine was only available in Alexandria and Cairo.

The peaceful takeover of Cyprus in 1878–80 produced a disgracefully large number of deaths from malaria. Had the British Foreign and War Offices included in their planning a visit to a library they would have discovered that the centre of the island contains a plain, at one time drained by a river which flowed into Famagusta Bay. However, in summer the river now never reached the sea but petered out in marshland, a classic environment for the malarial mosquito. After the Venetian surrender of Cyprus to the Turks in 1571 trees were cut down, the land went undrained, sugar and rice plantations were left to degenerate into marshes, and the population declined by more than half. Seventeenth-century travellers to the Holy Land usually inserted a clause in their contract to prevent dalliance 'in any port in Cyprus', so clearly it had become notorious quite soon after the Turkish takeover. Cyprus was perhaps the worst example of malaria-by-neglect in the whole Mediterranean.

22 The discovery of avian malaria, like the discovery of avian TB, was of great value in the early days of the study of each disease. It is of primary importance, long before a 'cure' is discovered, to trace the habits and characteristics of a disease. The alternative, which is not unlike throwing coins into a fountain, is to apply to a mortally diseased person a series of potential cures in the hope that the disease will disappear before the patient dies. This approach is still more common than most medical researchers would admit.

23 German expertise continues to excite admiration. In West Germany IG Farben, split into three parts, can be counted in the lead in organic chemistry. Even in the East, which received its mathematically correct share of IG Farben's assets in 1945, the Germans lead their fellow Communists in Comecon in terms of chemical production. Compared with the West Germans, however, except in terms of war gases they are twenty years behind.

24 Pamaquin was to be followed in 1930 by the more successful metabrine, which was copied by both the British and Americans and became the atabrine of World War II. Atabrine produced many side-effects, most of them unpleasant: it turned the skin yellow, was toxic in large doses, and was capable of killing children. Though atabrine was not as good as quinine, its production could nevertheless be expanded rapidly in the three countries which made it: Germany, which supplied Italy and occupied Europe; the United States, which had to fight all through the Pacific, taking malaria (and the drug) wherever its forces went; and Britain, which needed thousands of tons for India and the Middle East.

Another drug, chloroquine, was invented by the Germans in 1934 and perfected by the Allies, who successfully interpreted the German chemical patent and modified manufacture so that it was available as an alternative to metabrine. Chloroquine is a much better suppressive than quinine or metabrine, and need only be taken once or twice a week instead of daily.

The next drug was a British invention, proguanil, produced at the end of the war, which became a standard daily prophylactic for years afterwards. It has perhaps the widest safety margin of all these drugs, and as rare an incidence of side-effects as quinine, until then the safest of all.

25   The malaria–quinine–blackwater fever syndrome is still not an absolute certainty, but a full explanation can be found in Bruce-Chwatt, *Essential Malariology*, Heinemann, 1980. A series of attacks of *falciparum* malaria destabilizes the equilibrium of the blood, in particular the haemoglobin count. The bile, liver and kidneys are affected, and the urine is dark red or black, as is the vomit; hence the name. It was confused for many years with both yellow fever and 'bilious remittent' malaria.

When quinine is administered to an individual suffering from *falciparum* malaria, it may precipitate an attack of blackwater fever, which can prove fatal. A shock can also perform this task – typically, a bad car accident. Before the days of synthetics, however, the choice was not wholly between malaria and quinine, and therefore blackwater fever. It was in blackwater areas, the malaria *falciparum* areas, that environmental control measures (drainage, insecticide etc.) were most vigorously prosecuted as an alternative to the 'dangerous quinine'. Since the advent of the prophylactic synthetic drugs blackwater fever has become rare.

26   Immunity against the more severe form of malaria, *Plasmodium falciparum*, is due to the high incidence of Haemoglobin S (HbS) amongst negroes in *falciparum* areas. The disease becomes starved, effectively, of oxygen, because of what we now call sickle-cell anaemia. Drugs attempt this result synthetically.

The development of immunity in negroes, which has had such a vast influence on history, appears to have been due to a process of natural selection. The only original inhabitants of the 'white man's grave' of West Africa, for instance, were negroes with malarial immunity. Asians, whites or non-negro blacks do not have the ability to develop such immunity, at least during the timescale of the last five hundred years. Other forms of haemoglobin, such as HbC, HbF and HbE, exist in malarial areas of southeast Asia. There is no evidence either way that any protection is offered against *falciparum* malaria by these genetic variants. Induced immunity may ultimately prove more effective than other control measures.

27   Syphilis was the great curse of the New World upon the Old. It was incurable for three centuries, and then painfully and dangerously 'curable' only by the application of a varying amount of mercury, which either killed the patient or cured the first two phases of the disease. Before the use of induced malaria the tertiary phase needed potassium iodide. Salvarsan, or 606, invented by Ehrlich in 1910, was better than mercury, but did not replace potassium iodide or induced malaria. The first true cure came in the late 1940s with antibiotics.

28   The Tamil minority in Sri Lanka, related to the short, very dark natives of southeast India, were imported to replace the indigenous Sinhalese who were regarded as 'poor workers'. The Tamils are now a very strong minority and feel aggrieved about their lack of civil rights, economic advantage and freedom of expression. They seek local autonomy, a state within a state, which is denied them; hence violence replaces political action.

29 'Work ethic' has nothing to do with religion or guilt, the first half of the twentieth century notwithstanding. No subsistence population in easy circumstances, such as in traditional Tahiti, needed to 'work'. The white man did not invent work, but he brought to the post-Renaissance world higher populations, market ideas, specialization, land tenure, and generations of incentives. Which of these influences, physical or mental, were paramount cannot be answered by slogan chanting, but consider modern Chinese agriculture, which has had the highest growth rate in the world (12 per cent annum compound) since private farming was permitted (*Economist*, 2 February 1985).

30 At least 20 million people throughout the world became 'indentured workers'. Except in the case of a small minority of true negroes, this was only possible thanks to the existence of quinine. There is considerable evidence that the indentured worker was preferred, not only because of the low pay he could be offered, but also because of the undistracting conditions of his life. In Sri Lanka even today many barracks for the Tamils have no artificial light of any sort, leaving little for the indentured worker to do except work, eat, sleep and of course breed. It is quinine which has led to the massive minorities of imported people, not only in Sri Lanka but also in Guyana, Fiji and most of the islands of the Indian Ocean.

31 Probably as much quinine is used today as ever before, but in soft drinks, not in medicine. Some tonic waters contain thirty parts per million of quinine. Does this mean active febrifuge, or bark, and if so, which bark? (There is also a non-cinchona bark, from the cuprea tree, *Remijia*; it is far cheaper than cinchona, and has a strong flavour, but only contains ½-2 per cent of active quinine.)

Other soft drinks, such as bitter orange, bitter lemon and the various colas, also contain quinine. Heavy consumption of such quininated drinks lowers body fever and reduces cramps of all kinds. Many women have noted that gin and tonic alleviates premenstrual tension. It is of course, the quinine which helps the condition, though the gin may promote the dispersal of the drug through the nerves and muscles of the sufferer.

The earliest recorded lemonade containing quinine appears to have been produced in New Orleans in 1843; Schweppes started volume manufacture of Indian tonic water in the 1870s, and of bitter lemon in 1957. Some claim 'quinine water' to be a British-Indian development, as a vehicle for the daily dose of febrifuge; others liked the bitter taste. French and Italian aperitifs also contain 'quinquina'.

Quinine was widely used in cold and influenza medicines until World War II, largely to reduce body fever. It is one of the safest antipyretics known, bringing a fever down by as much as 3°F; it is also a painkiller and dulls jangled nerves.

As late as 1947, the author saw bottles marked 'Not to be taken by expectant mothers' on sale in Kansas drugstores; the appeal of this elixir was that it could bring about miscarriage. It succeeded in perhaps 50 per cent of cases, but at a cost. The heavy dose of quinine resulted in very low blood pressure, slowed the pulse, probably induced depression, and sometimes led to permanent deafness with ringing in the ears (tinnitus). But Old Wives would always prefer quinine to a back-street abortionist; quinine was safe up till about the fifteenth week of pregnancy. These mixtures may also have been available in other states of the USA, and in other countries.

32 In August 1985 the London *Times* reported on an article in *Science* (vol. 228, no. 4703), confirming that the Chinese, engaged in just such a search, have found a quinine substitute derived by low-temperature extraction from *Artemisia annua*.

Sugar

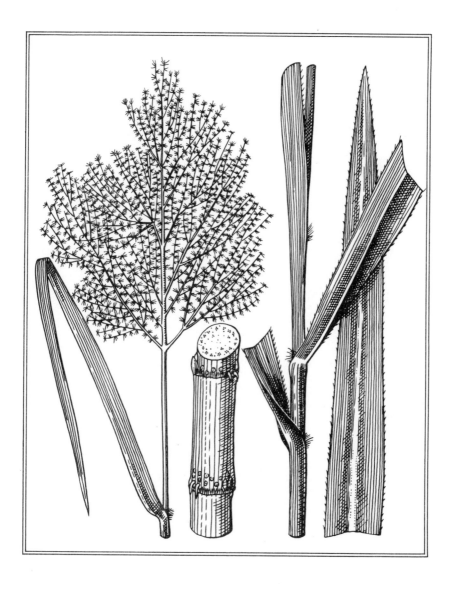

Sugar
and the Slave Trade

The ingenious wording of a certain English china warehouse's advertisement for sugar basins in the early 1800s exploited the contemporary wave of liberal thinking: 'East India Sugar not made by Slaves,' the pots were printed, thus enabling the purchaser to display his conscience publicly. 'A Family that uses 5lb of Sugar a Week', the advertisement continued, 'will, by using East India instead of West India, for 21 Months, prevent the Slavery, or Murder, of one Fellow-Creature! Eight such Families in 19½ years will prevent the Slavery, or Murder of 100!!'[1] The equation of 5lb for 21 months, or 450 lb being equal to the life of one slave, was a very extreme calculation. Most of the evidence from the seventeenth century, when conditions were primitive, life was cheap, and slaves could be obtained relatively easily in West Africa, would equate one life with half a ton of sugar. By 1700 it was parity: one ton = one life. By the end of the eighteenth century, it was nearer 2 tons equalling one slave's life. So these figures are polemical rather than accurate. Yet this is the central conundrum of the whole sad story; it is also one of the major puzzles of modern history. Sugar remains one of the great moral mysteries.

It was, and is, in absolute terms, a not especially cheap source of human energy. In the eighteenth century it was much more expensive in real terms than cereals. Before the sixteenth century the whole of the European world had managed with minuscule quantities of sugar, a mere pinch per head for the whole of history. The glories of the Renaissance were created on the basis of a teaspoonful per head of sugar per year. Sugar is unnecessary to any endeavour, but it is addictive. In the years 1690–1790 Europe imported 12 million tons, which cost, in all, about the same number of lives. Today, Europe's consumption is well over 12 million tons a year, and there are no slaves except consumers.

Sugar cane[2] is native to Polynesia, where it was invested with near-magical properties, a mythology arising, perhaps, from the fact that small pieces were often found washed up on foreign shores where they readily flourished. This is the probable explanation of its movement to China, India and elsewhere. Sugar cane was widely used in ancient India, and in China was chewed as an aphrodisiac sweetmeat in about 1000 BC. But it was first refined into sugar as such in India some three hundred years later, at Bihar on the Ganges, and thence introduced as sugar to China.

Indian sugar was made from a variety of cane called *puri* and it was this variety that spread slowly westward for the next two thousand years, to be improved in the eighteenth century by strains from Polynesia and Indonesia. Columbus is known to have introduced *puri* to Haiti from the Canaries in 1494, and the name was naturalized in the British West Indies as *creole*.[3] Europe did not see sugar until the Middle Ages, when it first arrived in the Mediterranean. It was, however, once believed that Alexander the Great had come across sugar cane in the Indus valley in 325 BC.

Long before sugar cane was distilled and crystallized, honey was the great sweetener – the bee is a very efficient sugar concentrator. The first dated reference to honey bees is in Egypt in 5551 BC, and there are many in Babylonian sources and throughout the Old Testament. In Egypt magical properties were attributed to honey: it was made into a syrup which was supposed to prolong active life in the aged; it was an ambrosia which was meant to tranquillize; and it was used as an aphrodisiac. It played a part in all sorts of ceremonies, both sacred and profane, in Ancient Egypt, Babylon, Ur, Persia and India. Not surprisingly, Moses forbade the ceremonial use of honey, since it had acquired from the Egyptians the carnal overtones associated with the abuse of alcohol.

In no ancient civilization before about 650 BC is there any evidence of bee husbandry, defined by Virgil, among others, as control of swarming. Usually all honey was 'hunted', and stolen from wild bees. In Homeric Greece the process of honey-making was not completely understood, and the character of the end product was attributed more to the quality of the bees than to what the bees fed on. In Roman times Cato, Pliny the Elder, Varro and above all Virgil described bee-keeping in a domesticated sense for the first time in history. Virgil's Fourth Georgic contained the first verses on bee-keeping and it became known throughout the civilized world. In consequence apiculture, as opposed to honey-hunting, spread throughout the Mediterranean. Despite Moses, the early Christian Church associated honey with magical properties – sacred this time, as opposed to those which the Ancient Egyptians had claimed for it. Honey was used in the rite of baptism until about AD 600, and the bee was credited with the virtue of virginity – hence the obligatory use of beeswax candles in Catholic churches.[4]

Long before they became Christians, the Celts, Germans and Slavs used honey to make mead. Throughout northern Europe, from the Urals to Ireland, honey and mead formed part of the diet of the more privileged, and mead production in Bavaria and Bohemia and on the Baltic coast reached industrial proportions in the Middle Ages. But in the early fifteenth century honey began to be supplanted by the new cane sugar, and so mead gave way to beer. In Russia this change did not take place until the arrival of beet sugar in the nineteenth century. Even as late as the 1860s an industry of bees–honey–mead existed in parts of Russia, and Tolstoy mentions it in a letter to his wife.

There seems to be a clear connection between weather and the sweet tooth. Countries with a vine-growing climate were always much more modest consumers of sugar or honey than were those countries which could not produce wine. What weight must be given to the sunshine and to the sugar in the wine, no one can say, but sugar/honey and alcohol were quite clearly alternatives in all cultures in the post-Roman world, long before Dorothy Parker said the last word on the physiology of alcohol dependence: 'Candy's Dandy, but Liquor's Quicker'.

Mohammed banned alcohol to his followers. At first alternative stimulating drinks in Moslem countries had to depend upon honey for sweetness, but within a hundred years of the Prophet's death in 632 sugar cane and sugar production had been introduced from Persia and were established in Syria, Palestine, the Dodecanese, Egypt, Cyprus, Crete, Sicily, North Africa and southern Spain. The sugar industry survived the gradual expulsion of the Moors from the Mediterranean littoral, and was carried on by both Moslems and Christians as a profitable, expanding concern for two hundred years from about 1300.[5] The trade (as opposed to production) was under the dominance of the merchant bankers of Italy, with Venice ultimately controlling distribution throughout the then known world. The first sugar reached England in 1319, Denmark in 1374 and Sweden in 1390. It was an expensive novelty,[6] and useful in medicine, being unsurpassed for making palatable the odious mixtures of therapeutic herbs, entrails and other substances of the medieval pharmacopoeia. Its price was far too high for it to become a normally consumed food:

Price of 10lb of sugar or honey, expressed as a percentage of 1 oz of gold (average of London, Paris and Amsterdam)

Period	Sugar	Honey
	%	%
1350–1400	35	3.3
1400–1450	24.5	2.05
1450–1500	19	1.5
1500–1550	8.7	1.2

In two hundred years the price of both sugar and honey declined dramatically. There is evidence that both reductions were due to increased production in the cane industry.

These were the years when the first sugar from canes grown outside the Mediterranean became available on the European market. The Portuguese planted canes in Madeira, the Azores and Sao Tomé, and the Spanish in the Canaries, half a century or more before there was any production by the Spanish in the Caribbean, and it was they who first used slaves outside Europe. Until 1550 the only sugar imported from the Western Hemisphere consisted of a few loaves brought as proof of the possibility of production, or as mere curiosities. The West Atlantic islands and the New World had no effect on production, distribution or prices until the latter half of the sixteenth century, and only became dominant from about 1650. Before 1600 Venice gave way to Amsterdam as the great entrepot for the sugar trade, as for the spice trade, and for the same complicated reasons. All this, as we shall see, was part of the movement of the epicentre of world trade out of the Mediterranean and towards the Atlantic.[7]

So this chapter is about the other end of the sugar story: how an unnecessary 'food' became responsible for the Africanization of the Caribbean. It concerns only a very small part of the world, but one which, until 1800, was responsible for more than 80 per cent of both sugar and the trade in slaves. As a direct result, this small region was simultaneously responsible for nearly half of all the seagoing effort, naval and civil, of the western European nations. The story must be allowed to tell itself, but it would be worth asking first why people came to eat refined sugar at all, except as some sort of curiosity, and why they became addicted to it.

What exactly is sugar, biochemically? All edible plants contain, in varying proportions, fibre, protein, fat, starch and sugar. All vegetarian and omnivorous animals, including man, convert fibre and starch into sugar by biochemical means. Sugar is then made available in the bloodstream as a source of energy. Starch and sugar occur in all fruits and vegetables, and, before the arrival of industrial cane and beet sugar, mankind managed well enough without refined sugar, which is pure, or nearly pure sucrose; the form of sugar contained in fruits and vegetables is fructose.

When pure sucrose is consumed in large quantities, the metabolism of the whole system is altered. If a person eats a fruit containing, say, 10 per cent fructose and 10 per cent glucose, the remaining 80 per cent of dry matter has to go through a number of digestive processes to make the sugars available. In consuming pure white industrial sugar, whether from cane or beet, the stomach has less work to do, and energy is produced and used up in a sudden flood rather than as a steady drip.

When a large quantity of sugar is consumed (and some people eat and/or drink over 4lb a week) it can meet the whole of the body's energy requirements, and the rest of the food or drink consumed becomes a mere vehicle. The production of starch- and sugar-converting enzymes is inhibited once the body habitually gets its sugar requirements directly from sucrose, so the stomach finds it difficult to digest any accompanying starch or fibre. Naturally people will avoid foods which they find indigestible, so food manufacturers then reduce the fibre content in processed or packaged foods. A vicious circle is created in which the victim becomes hooked on a constant flow of industrial sugar to the bloodstream and cuts down on fibre.

The white sugar addict becomes liable to obesity, tooth problems and malnutrition; the last leads in extreme cases to the kind of 'crowding out' which can cause vitamin and mineral deficiency problems and probably even cancer of the intestine. Because of the speed with which white sugar becomes available to the metabolism, the addict's blood-sugar level rises and falls very rapidly as the pancreas works excessively hard to deal with high inputs of sucrose to the stomach. The body becomes used to a feast/famine syndrome in the blood sugar, and this produces an addiction which is chemical, not psychological. The bloodstream signals a deficiency which is self-induced, and the whole cycle may be repeated within an hour. A true sugar addict cannot do without some kind of reinforcement at very frequent intervals.

In England, where heavy consumption of white sugar arose earlier than in any other country, the preference for white bread also began as a result of sugar addiction. Social historians have had a field day with the Englishman's illogical preference for white bread, which began at the end of the eighteenth century, and there has been much play with that well-known obsession of journalists, politicians and other commentators, the British class system. 'Class' has been held to be responsible for everything from scientific and technological failure to the decline of regional cookery, with side-swipes at clothing, pop and sex on the way. White bread is no exception. But there is a very good biochemical reason for this unique English preference which would still hold true even if the English public school had never been invented. In the face of high consumption of white sugar, the enzymes required to digest wholemeal bread are absent, since they are literally killed by industrial sugar. Conversely, if one eats enough dietary fibre, one does not crave sugar. If one eats both fibre and sugar, as in some breakfast foods, one gains an excessive amount of weight, and the sugar negates any good that the fibre might do.

In 1800 the United Kingdom consumed more than 18 lb of sugar per head per year. As sugar then cost more than five times the price of energy

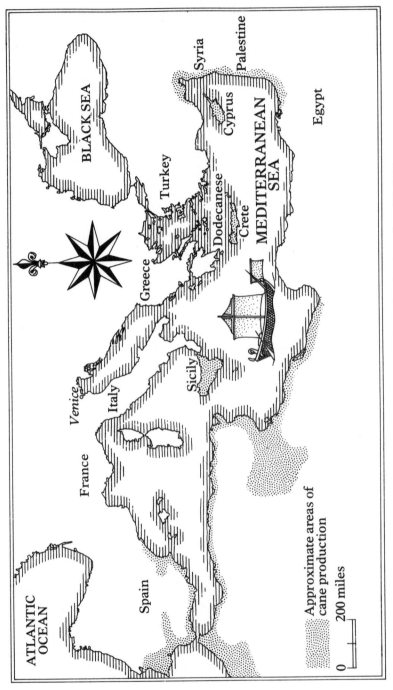

Arab Developments in Sugar Cane Production after AD 700

in the form of flour, and at least ten times the price of energy in the form of potatoes, only the rich could afford sugar. So some people must have eaten double the average – say 36 lb per year – which is more than 11oz per week. At that rate of sugar intake, few could digest high-fibre bread. So the White Bread Myth became established for the rich by biochemistry, class warriors notwithstanding.

Sugar addiction is often considered an adolescent problem to be put aside later. Yet it is usually replaced by addiction to alcohol, which enters the bloodstream even more quickly than sugar, almost within minutes if unaccompanied by food. There is an absolute correlation between high sugar consumption and high alcohol intake outside meals. Both sugared drinks and alcohol outside meals give a 'lift' to the blood sugar which humans on a balanced diet (including wine with meals) do not require.

What the presence of sugar, and the parallel absence of roughage, in the diet is doing to affluent Westerners has been discovered in the past generation; the effects include at least one form of cancer due to a low-fibre diet. The story is now so well known that there is little need to repeat that sugar, after the illegal drugs, and tobacco and alcohol, is the most damaging addictive substance consumed by rich, white mankind.

In the ancient world slavery accounted for perhaps two-thirds of the population of Athens under Pericles, and perhaps half the population of what is now Italy under Julius Caesar. Ancient Egypt probably had a higher percentage. In Ancient Greece and Rome the slave population was increased by servile births, by prisoners of war and by debtors. Slaves included house servants, athletes, doctors, accountants, artists, philosophers, show-business personalities, and men and women used for the pleasures of the flesh. Almost all the professions – with the exception of the army, the priesthood and the law – were manned by slaves, and many by slaves alone. Priests, farmers, soldiers, legislators and some of the traders and artisans were citizens. To a poor man in Athens or Rome, to become a slave might be a way to eat, to live in a better house, to avoid debt and to be secure under the protection of a good master. Until the insane cruelties of the emperors who succeeded Julius Caesar, there is no substantial evidence that slaves were treated worse in Rome than in Athens. But after about 50 BC servitude became increasingly onerous. Slaves had no civic rights; slaves could only give evidence under torture (because without torture, the witness could not be believed): when a man died suddenly all his slaves were murdered, regardless of guilt or innocence; industrial slavery on a vast scale was set up in Sicily, Spain and the Po valley. Slavery became a structured part of the institutional sadism of the later Roman State, and of a moral deficiency which casts doubt on any admiration of Imperial Rome.[8]

The Arabs, by and large, understood the inefficiencies of servitude. Moslem thought was much more commonsensical than the contradictory Romano-Christian views on slavery. Though Christianity had been adopted first by the poor and the slaves of the Mediterranean world, the Church was ambivalent about slavery once Christianity became the established religion. For slavery was essential to civilized life unless the rich and powerful were prepared to work themselves. The well-born could fight and write verses and philosophize, appear in law courts and engage in disputation, haggle and trade, gamble and socialize, but not for them the hard graft of the same daily grind of 'business'. This situation could be sustained only in a sedentary, settled society, and slavery diminished with the break-up of the Empire, to be replaced by warfare, murder, the slaughter of prisoners and all the other horrors associated with the time between the end of the Roman Empire and the onset of feudalism.

Feudalism[9] grew in Europe quite logically after the troubled times associated with the end of Roman rule. After the barbarian invasions, settlements were always threatened by strong and ruthless peoples wandering through the former Empire. A tribe in the Loire valley, for example, would seek the protection of a neighbouring lord and his gang of desperadoes. At best this protection, which was exchanged for so many days' work on the lord's land, or so many bushels of grain for the lord's grain store, or so many men-at-arms for the lord's company, was no more irksome than the protection which we all pay today in another form for defence from our foreign enemies and for crime-free streets. At worst it was no better than the protection paid to a mobster by a small store in any large modern city which has an inefficient or corrupt police force. In time, feudalism became institutionalized so that the king had a small number of great lords who supplied him with men-at-arms; the great lords had many minor lords, the minor lords had free men, and the free men either had to fight for the lords themselves or to produce goods and services provided by serfs.

Serfdom was better for the servile than slavery, since they could enjoy a home, marriage, a little land and some communal life; nor could the serf be separated from the land, though when the land was sold the serf went with it. By the time that the African was enslaved by the European, serfdom had succeeded slavery in most of Europe for nearly a thousand years, and except in Germany, Poland and Russia it had been much modified; in England, the Netherlands and parts of France and Portugal it had actually given way to a cash relationship between landlord and tenant. Even where the full rigours of medieval serfdom still obtained, such as in darkest Mother Russia, the serf was better off than most slaves.

The Arabs, too, found serfdom more efficient than slavery, and for the same reason. While the feudal lords might take from the peasant two or three days a week, the man had his own land to cultivate for the rest. The serf was not the passive victim of the master's incompetence, as were many slaves who suffered malnutrition or even starvation. The serf had protection and security and an incentive to work hard. The slave was insecure, with no remedy against an unjust master, and with no incentive except the lash to help him labour. The serf had the right, too, to protection from abuse and was allowed to worship at the church or mosque, whereas slaves could not enter the latter and were not encouraged in the former.[10]

Though the Arabs had and still have a very low opinion of actual physical work, they were excellent planners, managers and agriculturists. In addition, all Arabs enjoyed haggling and bargaining, and would trade in anything, including slaves. Arabs made warfare pay for itself and Moslem pirates made the whole maritime life of the Mediterranean insecure; Arabs sold prisoners of war; Arabs traded far down the coast of West Africa and raided the interior for the black servants and personal bodyguards much esteemed in Moslem cities, just as they established themselves in Zanzibar for the same purpose. But under the Arabs the negro slaves were few in number, generally house servants and rarely acquired for industrial or agricultural purposes.

In 1425 Prince Henry the Navigator of Portugal, a leader who, despite his sobriquet, rarely left his castle, established by proxy a Portuguese settlement in Madeira. The first sugar cane was pulped and refined in a plant on the island in 1432, the Europeans meanwhile having destroyed most of the island's woodland by accident and most of the island's natives by design. This industry was worked by more than a thousand men brought from Portugal, including convicts, debtors and stubborn Jews who refused to adopt Christianity. None of them was either slave or African.

Though Columbus had taken cuttings of sugar cane from the Canaries to Haiti in 1494, there is no record of successful culture, and the canes probably went wild. By 1510 there were perhaps a dozen sugar estates on the islands, using imported horses, imported machinery and imported workers – there was very little economic advantage in growing sugar in the West Indies at this time, since everything except land and fuel had to be brought in. But one technical point favoured the West Indies over the East Atlantic islands or the Mediterranean littoral, and that was the husbandry requirements of sugar cane.

Cane is a grass, and all grasses need fertile soil. Before European knowledge of plant nutrients had progressed beyond the rules codified in verse by Virgil and Ovid, the fertility problem could only be met by fallowing

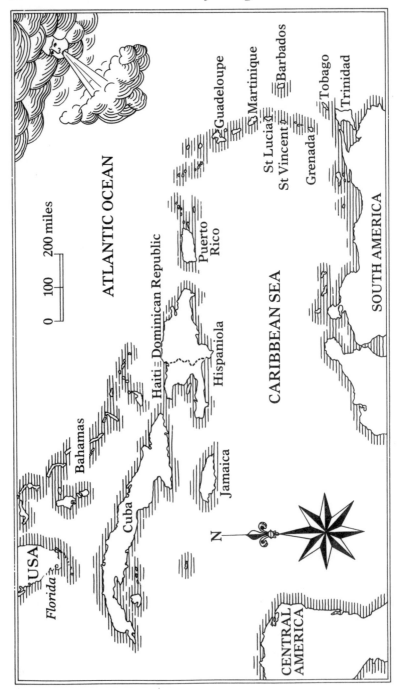

The West Indies

part of the land every second or third year, as was done for other crops in the Middle Ages. The islands, however, were originally well wooded and relatively underpopulated and so presented the opportunity to practise the age-old custom of *jubla*, or 'slash and burn', still followed in Africa and India today. This system involves burning an area of scrub or savanna, or even forest, growing one or more monocultural crops[11] on the cleared ground, and then moving on and repeating the process. For up to three years the residues of the burnt plant material produce very much better crops than the same land under continuous cultivation, even if fallowed every other year. In its ideal state *jubla* would involve burning a patch, cultivating it for a few years, and allowing perhaps twenty years for the scrub or forest to recover before starting the cycle once again.

Apart from the single bonus of the inherent fertility of forest land, there was no logical advantage in the early sixteenth century in growing sugar 4000 miles and three months away from the market in Europe. Prices in 1500 had been falling for 150 years, as we have seen, and for the next two generations they were to fall again, so that by 1560, and measured against gold, sugar was barely half the price in Europe that it had attained in 1500. Though prices would rise steeply after 1570, the quantities involved were still tiny in comparison to the trade today. It is certain that the whole of Europe's sugar production for the year 1600 could be contained in one modern bulk carrier: far less sugar than is eaten in one year in present-day New York City, London or Hong Kong.

Prince Henry the Navigator, perhaps obsessed with the myth of Atlantis, sent many ships along the coast of West Africa, as well as to the East Atlantic islands – the Canaries and the Azores. One of his ships, returning fruitlessly from Equatorial Africa in 1443, fell upon a galley and captured and enslaved the crew. These men, who were of mixed Arab–negro parentage and Moslems, claimed that they were of a proud race and unfit to be bondsmen. They argued forcefully that there were in the hinterland of Africa many heathen blacks, the children of Ham,[12] who made excellent slaves, and whom they could enslave in exchange for their freedom. Thus began the modern slave trade – not the transatlantic trade, which was yet to come, but its precursor, the trade between Africa and southern Europe.

The novel feature of the sugar slave trade was not only that the slaves were negroes and the traders were white, but that a whole new mythology grew up to justify the industry. The negroes were children of Ham, and therefore unworthy of consideration as human beings; free white men could not be expected to work in the sugar plantations; the negro was discouraged from becoming a Christian, and forbidden to read and write, so that he could be regarded as hardly human. These theories became accepted within two generations of the first shipload of slaves arriving in Lisbon in 1443, and they were perhaps necessary to blunt men's minds to this monstrous aberration in the history of the Western world. Long before Spanish Christians knew about the African homeland, the port of Seville was a thriving slave market into which a hundred shiploads of negroes were brought every year from the Portuguese trading stations in West Africa.

These negroes were only necessary to the agricultural economy of Spain because of sugar. The Spaniards, who were less able agriculturists than the Arabs, found that they grew less sugar with more labour. So they were obliged to invest heavily in slave labour. They also neglected the husbandry and the irrigation which the Arabs had carried out with the use of serf and free labour.

How much of the incompetence of the Renaissance Spaniards in Spain as well as in the colonies was due to poor husbandry, ineffective irrigation or slavery, no one can now tell. Did the institution of slavery lead to the neglect of husbandry, a virtue and an art which requires the personal attention of the proprietor? Did slavery lead to the decline in irrigation works, which require daily attention if they are to prove successful? Did slavery in the modern world bring into disrepute the whole idea of manual labour, as it had in Athens or Imperial Rome, states which in an earlier phase had been proud of their artisan and agricultural forebears?

Or was it a question of capital? If most of the capital were invested in slaves, there would be little left for irrigation. It should be remembered that before the eleventh century the Arabs had installed irrigation works in Algeria, Morocco and southern Spain which were unsurpassed until the twentieth. Those built in 980 in Marrakesh, for example, were still operational when the French arrived nine hundred years later. Slightly renewed, they are still working, but have required a constant investment of capital in repair and refurbishment, an investment impossible if the capital had been pre-empted by the high capital cost of slavery.

If the Spaniards were responsible for inventing the moral respectability imposed on African slavery, it was the Turks who made it an economic necessity. Between 1520 and 1570 the Ottoman Turks conquered Cyprus, Crete, the Aegean, Egypt and much of the North African littoral. In doing so they extinguished most of the Mediterranean sugar industry since, unlike their fellow Moslems, the Arabs, whom they replaced, the Turks were not great traders, and held the infidel in fierce, isolationist religious contempt. Sugar prices rose steeply after 1570, more than quadrupling, measured in real terms, in the last thirty years of the sixteenth century. Meanwhile, wealthy western Europeans had turned to sugar rather than

honey, and even before the advent of tea, coffee and cocoa sugar dependence was great enough to bring the New World into the reckoning to redress the balance of the Old. In the Europe of 1600, only Spain produced
sugar in quantity. The rise in the price of sugar over the last thirty years of
the sixteenth century was partially caused by inflation, which increased
the money supply throughout Europe.

There is much argument about gold and silver treasure from Latin
America and its effects on Europe. At this time the very unimaginative
system of management did not allow the treasure to be employed usefully
in Spain or the colonies, and this vast capital benefice, a windfall equivalent
to five or ten years' national income, was largely lost through ignorance
and the rigid habits of a monarchy in theory absolute, in fact limited in
ability. This affected the whole of Europe, not least Portugal which was a
dependency of Spain from 1580 onwards. By 1640, when Spain relin-
quished her dominion over the country, Portugal had lost for ever her pre-
mier position as an Atlantic trader, settler and merchant. Portugal was
overtaken by the Dutch, and the Dutch by the English and French, in the
sugar trade, in the colonies and in the slave trade itself.

In an astonishing burst of buccaneering enterprise, the Spanish conquered
in a few years, with only a few thousand men, most of what is now Latin
America, except for Brazil which was Portuguese. Spaniards were reluctant colonists: fewer than half a million of them emigrated to the Caribbean or to mainland America in the first century after Columbus, and
fewer still wanted to remain there, so that the government had to use land
grants to persuade some of them to settle permanently. It was the ambition
of every Spaniard to make enough money to become a don and buy an
estate back home in Spain. Some of them, for example the followers of
Cortes and Pizarro, did this in a very few years. Only a small number of
men were involved: 600 with Cortes, 180 with Pizarro; there was enough
Inca or Aztec treasure for all. But there was no comparable treasure in the
West Indies. The new owners found that they either had to work themselves, a fate they had crossed the Atlantic to avoid, or find someone else to
work for them. However, those few indigenous natives who had not been
destroyed by the settlers were hiding, shy and frightened, in the mountains, or actually eating the Spaniards (two landowners were eaten by
Caribs in Hispaniola – modern Haiti – in the 1520s).

Though African slaves had been imported to Cadiz and Seville by the
Portuguese since 1450, and many were at work in the sugar fields and rice
paddies of southern Spain, there was not a surplus to permit the required
export of twelve with each gentleman adventurer. Prices of slaves inevitably rose, and the gentlemen adventurers became deservedly unpopular.

Blacks had to be imported in quantity to replace those crossing the Atlantic. Ultimately, the obvious solution presented itself, and from about 1530 slaves were sent direct from Africa to the Caribbean.

In 1514 Bartolomé de Las Casas (1474–1566) was granted a block of land in the Spanish colony of Cuba; attached to his land were about a hundred native Amerindian Caribs. The attachment of native humans to land conquered by a successful army was a standard European procedure and an acknowledged part of the feudal system: witness, in the years following the battle of Hastings, the transfer of Anglo-Saxon serfs from English to Norman 'ownership', formalized in the Domesday Book. The Spanish had used this method to reward successful survivors of the battles to reconquer Spain itself from the Moors, then in the settlement of the East Atlantic islands, and then in the Americas. The grant of the people living on the land was known in Spanish as *ripartimento*.

In the Caribbean, the native Amerindian Arawaks had been superseded on most of the islands by the much more aggressive Caribs. The Arawaks were a short, squat people of gentle disposition, widely dispersed through what was to become Latin America, from southern Brazil and Bolivia to Florida and the Bahamas. Arawak means 'meal-eaters' and they were vegetarians, cassava being their staple food. They were skilful weavers and workers in stone and metal, including gold. The Arawaks were no match for the Aztecs, Incas and Caribs, all of whom displaced them, enslaved them or destroyed their social structure.

The Caribs were so named by the discoverer of Cuba, Columbus, from the Spanish *cariba*, meaning 'a valiant man'. They were cannibals, and gave the English language that word by corruption and as a pun, 'caribal' being transduced into 'cannibal' in allusion to the canine voracity of the race. They needed animal protein, refused vegetables, and probably came from the very middle of Brazil, migrating to the Caribbean in continual search of meat, which they preferred to be hunted, not domesticated. Like the Arawaks, they objected to their harsh treatment as slaves, and many of them either pined or contracted white man's diseases and subsequently died. It was after probably more than half the natives had died that in 1517 Las Casas suggested the African negro as an alternative.

That was only twenty-five years since Columbus' first expedition, in 1492, but much had been done in the meantime. That first expedition had discovered Watling Island, Rum Cay, Ferdinande, Crooked Island, Cuba and Hispaniola, and Columbus had brought back to the greedy merchants of Seville not only gold but bananas, cotton, parrots, curious weapons, mysterious dried and living plants and flowers, many dead birds and beasts never before seen in Europe, and five Indians, probably Arawaks, taken to Spain for baptism. Columbus returned to a hero's welcome, because, though he had discovered neither Japan nor China, neither the Philippines nor Indonesia – which was the object of the expedition – he had found the New World. By the time of his fourth voyage, in 1504, he was a dying man, but he had discovered most of the West Indies, initiated settlement in at least twenty islands, and already sent back to Spain enough gold to have paid all his expenses several times over.

Las Casas on his Cuban landholding was a good example of the muddled thinking which leads men to espouse the lesser evil. It was because the native Caribs and Arawaks were being forcibly made to work at jobs they could not or would not do, and because the majority showed every inclination to be killed, or to pine to death, rather than adopt slave status, and because in Peru and Mexico even worse treatment of the natives was considered quite normal, that Las Casas had suggested the introduction of blacks. They were known to be docile, apparently not to object to servility, and to work well and willingly. Thus began the transatlantic slave trade.

But Las Casas lived long enough to repent his choice. If the treatment of the native Indians all over Latin America was bad, the African slave trade turned out to be a great deal worse. Las Casas, the first priest to be ordained in the Western Hemisphere, became Bishop of Chiapa in Mexico and was known as the Apostle of the Andes on account of his work for the under-privileged Indians. Much of the trouble with both Indians and blacks arose because although the ruler of Spain, or a viceroy, would promulgate laws which were humane and decent, there was never anyone on the spot to make sure that they were put into practice. The most dreadful abuses became so commonplace that Las Casas resigned his bishopric and returned to Spain in 1547. By the following year he was conducting a nationwide campaign against the trade that he himself had started in order to save the Indians. His efforts failed but, if they had succeeded, the resulting law would probably have prevented the enslavement, transportation and early death of half the total number of Africans ever shipped. In 1554–5 Las Casas thought he had persuaded the Emperor Charles V to endorse emancipation. Instead, Charles was overcome with the need to save his own soul. He abdicated, and went to live in a little house next to the monastery of Yste, in Estremadura.[13]

It would not be the last time that the imperative need for private and personal salvation would contribute to the public condemnation of slavery and the slave trade. It is, however, a fairly rare event in history when the godfather of a new development has seen the error of his ways and tried to remedy his gross misjudgement. Las Casas was soon forgotten, and it would be two hundred years before slavery and the slave trade were as vigorously questioned and then successfully attacked. By then sugar was the most important commodity traded in the world.

The sugar trade multiplied at a compound rate of 5 per cent per year in the seventeenth century, by 7 per cent in the eighteenth and by nearly 10 per cent in the nineteenth. The inter-relationship of sugar, naval power, taxation, mercantile policy, capital investment and, above all, bondage and slavery is a complex one, perhaps best illustrated here by the story of three Caribbean islands, Barbados, Jamaica and Cuba, which together epitomize the developments in each of three centuries. None of the French islands has been included because, owing to English concern over the West Indies, five Anglo-French wars were fought in the eighteenth century and damaged the French sugar trade so much as to make French sugar production, distribution and exchange an activity promoted only during the intervals of peace. The sugar history of modern Haiti proves the point beyond argument.[14]

All the sugar colonies, of whatever nation, had a white-dominated pre-sugar history. Most of the colonies were established for mining, or to support mining, or for trade, or for piracy, or as the speculation of some individual feudal capitalist. Seventeenth-century Europe was dominated by religious conflict, by the Thirty Years' War, by the struggle between King and Parliament in England, by the decline of Spain and Portugal, and by the expansion and trading eminence of the Netherlands. These disturbances induced large numbers of dissidents of all types to settle in the New World and, in an age when life was uncertain, sea travel very dangerous and knowledge of the Western Hemisphere scanty, the pressures put upon emigrants must have been very high. Most of the emigrants from Europe were men, leading to an acute imbalance between the sexes in the colonies. Some of them must have stayed only a few years in each place before moving on, restlessly, to see if they could improve their luck in another colony; many must have ended their unrecorded lives in perhaps their third or fourth country of settlement. The numbers are not known for certain, and thousands of the more adventurous among them must have wandered all over the Caribbean without leaving any mark or trace of their peregrinations other than the travellers' tales related to sceptical companions.

An exploratory mission would find an uninhabited or little inhabited place. Upon the expedition's return to Europe, capitalists would be induced to subscribe and send their younger sons, or some other relative, to settle. The gentlemen adventurers would persuade men of the landless classes to accompany them, and often buy indentured servants, who were debtors or petty criminals, to settle with them. These indentured whites ('redlegs') were sentenced to seven years or more of bondage, following which they were free men, but debarred from returning to Europe. In such a manner was Barbados settled by the English, though Dutch, Spanish and Portuguese seamen were aware of its existence.

A virtually empty island the size of the Isle of Wight, or less than one-tenth the size of Long Island, Barbados had a relatively favourable climate, almost no native quadrupeds, an abundance of water and masses of timber. By the time of Charles II's Restoration, Barbados had become one of the most densely populated agricultural regions in the known world, supporting perhaps forty thousand people of whom more than two-thirds were white – a population density of 240 to the square mile.[15] In the decade 1660–70 Barbados was the greatest sugar-producer in the trade, but all the timber had been cut down and the soil had been exhausted – as a result it became the first island to import cattle in large numbers, in the hope that the land could be restored to fertility with cow dung. In the original settlements small proprietors had grown tobacco, indigo, cotton and ginger and other spices for export, and cassava, plantains, beans and corn for home consumption. When the land was worked out thousands of small proprietors left the island, and sugar, slavery and petty capitalism established the monoculture which was to be the common pattern all over the Caribbean. The first of the dissatisfied whites left to try their luck on other islands and in the Carolinas and Virginia, which at an early stage established a close relationship with the West Indies in general and Barbados in particular. The American connection involved the great and the good before and after the Revolution: Washington, Hamilton and Jefferson all owned property in Barbados.

Alone of all the West Indies, Barbados never changed hands in the interminable wars which went on until 1815. But already in 1670 there were fears that the white proprietors had become too few for defence. Proprietors owning more than 100 acres numbered nearly 16,000 in 1643. Because of land exhaustion and emigration, this class fell to 5000 in 1670. In addition to the white proprietors, there were over 30,000 indentured bondsmen and black slaves. These were better treated in Barbados than elsewhere since the island was always, even after monoculture was established, a place of smallish estates. The average in the late seventeenth century was about 200 acres, each with its own sugar mill driven by water, wind or oxen. But the population had changed drastically. In 1645, according to one account, there were 18,000 whites, of whom only 7000 were free, and only 4000 blacks, all of them slaves. The slaves despised the white bondsmen, many of whom had a nastier life than the blacks, since the purchase of their services was cheaper than the cost of a black man's labour. By 1675, blacks at 32,000 outnumbered the whites at 21,000, of whom less than half were free men.

The ratio of black to white in Barbados was never as wide as in other islands, however, and does not seem to have exceeded five blacks to one white. In some of the islands the proportion was 15:1 or 20:1. But how the census was taken when the population was as mobile as it was, and the next island was only hours away by small boat, no one today can tell. Barbados also experienced fewer slave revolts than other islands.

The small size of the Barbadan estates meant that many proprietors were actually working farms not much bigger than similar holdings in England. Any crop except sugar could have been grown without slave labour, and indeed white men did so before sugar entered the island's economy. Sugar was then, of course, the supreme cash crop. There was no demand in the eighteenth century for cotton; tobacco was more efficiently grown in Virginia; coffee and cocoa far less profitable than either. Except for gold, sugar was the only colonial product before 1750 which showed a trade balance in favour of the colony.

It was the boredom and hard work of sugar cultivation twice a year, at planting and at harvest, which made black slavery 'inevitable'. Cane planting was done by clearing a pit 3 feet square and a few inches deep, into which the young plant was dibbled. The object of the pit was to make subsequent weeding much easier. To save manual labour by ploughing was said to be impossible. If digging these pits in the hot sun was hard work, too much for whites, then harvesting the cane, crushing it and boiling the sugar was out of the question. The canes were crushed in mills, and the sugar then boiled out of the cane in a series of open vats in a sugar house. Refining sugar is similar to refining oil, the heavier and blacker fractions coming off first, then the whiter and finer ones. Today, every grade of sugar can be obtained from cane sugar – from molasses through black sugar, the various browns and then the finer whites. In the seventeenth century a small, primitive, on-farm mill would only produce molasses and one grade of sugar. The heat was fierce, since there was no means of cooling the sugar house. Temperatures of 140°F were recorded and, even at night, the temperature near the vat would be well over 120°F. Humidity would also be very high and therefore exhausting. It was a job for blacks, not whites: slaves, not free men.

After the timber had been cut down, around 1680, the Barbadans imported timber to fuel the vats from other islands, from mainland Guiana and from the Carolinas; they even imported coal from Newcastle in England. Even with imported fuel, there was little incentive to refine sugar beyond the necessary work to make it exportable, for the conditions in the fields and in the sugar house at harvest time were such that no white man would do the work, and no black man would work without the lash. Or so it was said. Thus came about the brutalization of slave and master alike.

Barbados offers a continuous statistical picture from 1637 through 1808, when the slave trade came to an end, until 1834, when the slaves were emancipated. A kindly island, it suffers no extremes of climate apart from the occasional hurricane; it remained in British hands continuously for more than three hundred years; there were fewer than 'normal' revolts; slaves were relatively well treated; and the figures are more accurate than in most areas of the New World. Barbados stands out among the sugar islands because of the absence of any apparent evidence of gross misery in the past. Yet Barbados had to import, legally, 350,000 slaves in a period of 175 years, plus nearly 100,000 white indentured servants; to these must be added whatever children were born during the lifetime of the estimated 100,000 female blacks who lived, at one time or another, in Barbados. If we allow one birth per female (surely not excessive), about 550,000 slaves, white and black, old and young, male and female, lived and died in Barbados in those 175 years. More than 40,000 indentured white and black slaves existed in 1675. Only 66,000, all black, were emancipated in 1834.

Strenuous efforts have been made in recent decades to work out the numbers involved in the whole transatlantic slave trade. Deer in 1949 gave a total of 12–13 million actually imported into the Americas. The consensus today is for less: the apparent figures are 11.7 million exported and 9.8 million imported into the New World between 1450 and 1900. This all-embracing figure does not include two further kinds of losses. The first, which must be set against all forms of the slave trade, consists of the losses ashore in Africa, of which the numbers are unknown. There were losses caused by the warfare, raids and ambushes instigated to produce prisoners to enslave; there were losses sustained during trafficking, marketing and transport; there were losses at the barracoon (or stockade) at the port of export; and there were losses as a result of disease, injury, attempts at escape and so on. The second, unique to the sugar industry, consists of the losses, unknown but obviously large, of the infants not surviving. When, after the abolition of the slave trade in the USA, the nineteenth-century cotton industry found itself short of workers, and slaves were energetically encouraged to breed, the slave population, with minimum imports, increased by ten times in sixty years – twice the free rate. This was despite a shorter expectation of life for slaves than for free women.

Depression among slaves on most estates lowered their animal spirits. Subjugated people have a low conception rate, and there was in any case an imbalance between genders, with males outnumbering females. Young slaves could do little real work until their teens, and it was cheaper to buy an adult slave than to debit the enterprise with a dozen years of keep for the young. The survival of infants born to slaves, therefore, was very much subject to the mood of the parents, who could see little hope for their children and did not strive officiously to keep them alive. Mortality before the first birthday was 80 per cent on some estates. There was little incentive to breed, both from the slave's and the master's point of view; indeed, deliberate breeding of slaves was considered by some Christian masters to be more wicked than the horrors of the slave trade. So, by accident or design, breeding was as discouraged in the sugar industry as it was later encouraged in the cotton kingdom.

In a notional plantation containing fifty slaves, five adults would need to be replaced every year just to maintain that number. Twenty per cent of those shipped from Africa died on board ship, so in order to end up with five live slaves in Barbados between six and seven would have to be shipped. There were further, unquantifiable, losses incurred on the long march from the interior to the port in Africa, and among babies stillborn, aborted or unreared in Barbados.

Between 1637, the date of the first sugar planted in Barbados, and 1808, when the last slave was legally landed in the West Indies, the value of a slave in the Caribbean varied between that of half a ton of sugar before 1700 and 2 tons in 1805. The average in the eighteenth century was about a ton per slave's life: 2 tons just before abolition of the slave trade. Two tons of sugar is less than a thousand modern schoolchildren would consume in a week in junk food, soft drinks and icecream.

For a very long time the average slave only produced one-tenth of his value each year. So one ton represented the lifetime sugar production of one slave who had been captured, manacled, marched to the African coast, penned like a pig to await a buyer, sold, chained again on board ship, sold on the island market and then naturalized to the conditions of the Caribbean ('seasoned') before he showed any profit to the plantation owner. The slave, bewildered if surviving at all, would have seen the matter in rather a different light. But few slaves would know that a black was worth just about the same as one ton of refined sugar, not per hour, not per week, but for the whole of his life.

Sugar is a substance which we now know that we can well do without, even today when it is cheap and freely available. Why, when its use caused so much death, cruelty and misery, did sugar move from a luxury afforded and used by a few in 1600, to a necessity for many two hundred years later? For every ton consumed in 1600, 10 tons were consumed in England in 1700 and 150 tons in 1800. In 1600, little of the sugar was slave-grown and none came from the West Indies directly to England. In 1800, nearly every ton of sugar imported into England was grown and harvested by slaves, and the ratio was one black man's life to 2 tons of sugar. In 1801 the population of England was about 9 million, and the consumption of sugar not less than 17 lb per head per year. This gives a total consumption in England of over 70,000 tons of sugar in that year. That was equivalent to twice the number (35,000-plus) of black slaves consumed in the islands in the production of sugar. On average, for every 250 English men, women and children, a black died every year.

This is the central social problem. Why did a relatively advanced society become so dependent on sugar as to allow such a slaughterous addiction? The sugar addiction in 1801, wherever it existed, killed proportionately more people than the drug trade does today. The drug trade differs, of course, in that it kills those hooked on the product, while the sugar trade mostly killed slaves.

Sugar, then, is the most notable addiction in history that killed not the consumer but the producer. Every ton represented a life. Every teaspoonful represented six days of a slave's life. Put that way, would anyone in eighteenth-century England have touched it? But, of course, few people in the eighteenth century did put the problem quite like that. It was argued then, after the slave trade and slavery had become such a major part of European life that it was impossible to remain neutral on the subject, that the life of the black in Africa was marginally worse than that of the black slave. Warfare, starvation and African slavery itself made it likely that the black's expectation of life in his country of origin was no higher than that of the black slave in white ownership. (This argument is morally akin to that of the hunter who says that as the prey will die a horrible death anyway, no ethical harm results from hunting the animal.) What these specious forms of apologia for slavery and the slave trade failed to recognize was that the deleterious effect upon the slave-owners and other beneficiaries of slavery was probably morally more damaging than any harm which, on balance, the black might suffer from being brutally removed from Africa; from the horrors of the Middle Passage; and from the degradation of the market and 'seasoning' in the West Indies.[16]

Sugar production also created demands for a return to Roman mass slavery. While Mediterranean slavery, both Moslem and Christian, continued up to and including the sixteenth century, it was of a particular nature. Slaves, like gold, jewels, works of art and wine, were luxuries, not necessities; they were part of what we would now call 'conspicuous consumption'. Slaves were paraded, displayed, bought and sold, given and received as objects indicating status. Their economic function was not much greater than that of any other ornament.

Sugar slavery was of quite a different order. It was the first time since the Roman *latifundia* that mass slavery had been used to grow a crop for trade (not subsistence) in a big way. It was also the first time in history that one race had been uniquely selected for a servile role: Spain and Portugal voluntarily abjured the enslavement of East Indian, Chinese, Japanese or European slaves to work in the Americas. They also made considerable efforts to end Amerindian slavery. It matters little that there were excellent if different reasons for these decisive actions. Experience led other Europeans to the same conclusions.

The saddest point about sugar slavery was that it was probably unnecessary. By 1600 it was within the capabilities of Englishmen and Dutchmen to use oxen rather than men, brain rather than brawn, and to adopt 'feudal' share-cropping rather than slavery. But in 1600 feudalism was considered old-fashioned by the Dutch and English, just as it was in 1800 by the new Americans. So, instead of share-cropping, both sugar and cotton gave rise to mass slavery. The slave population in the West Indian cane-brakes, and later that in the American cotton plantations, were the first in history to exceed the figure of 2 million in the Roman Empire at about 100 AD. [17]

But we should not, in our cosseted, sterilized, medicated twentieth century, forget that life for probably the majority of white men was also nasty, brutish and short. For seamen, the lash was commonplace. For all except the privileged, hunger, or at least uncertainty about food supplies, was the norm. Disease could strike rich and poor alike at any moment, usually with no possibility of prevention or certainty of cure.

Expectation of life, class by class, income group by income group, occupation by occupation, cannot be precisely established before about 1850; but there is no evidence that there was an immensely greater expectation of life in Europe than in the West Indies or Africa. The margin might be of the order of 25 per cent at any age: certainly no more. On the other hand, for those interested in the quality of life rather than its duration, there is also no comparison. The white man might be badly treated by circumstance, or his fellow men, or disease, but he had expectations, and it was the removal of expectations, however modest, however unlikely, however faint, which ruined the life of the slave.

The introduction of coffee, tea and cocoa into Europe progressively boosted the demand for sugar. Chocolate-drinking, coffee houses and afternoon tea all acquired a gentility far removed from ale-house bawdiness, and became first a luxurious amenity, then by the fourth quarter of the seventeenth century a middle-class necessity. But all three were crude and unconsumable, it was said, without sugar. From about 1680 the fashion for these hot drinks became a potent factor in the surge in sugar demand and consequent increased production, which progressively raised the sugar trade to the point of importance which it assumed in 1700.

During the second half of the eighteenth century the temperance cause developed into an important social movement, initiated by various Protestant denominations and therefore strongest in northern Europe, in countries such as Britain and the Netherlands. Sugared tea became the respectable alternative to beer or wine long before water was safe to drink without boiling. These changes in social habits significantly increased the demand for sugar and were probably responsible for about half of the increased trade. By the time sugar had become 'the indispensible companion of tea',[18] Jamaica had overtaken Barbados as the most important English Caribbean island.

Jamaica is a large island, more than twenty-five times the size of Barbados. It was discovered by Columbus in 1494, settled by the Spaniards in 1500, and then neglected for a century and a half. Like other islands, it remained a historical enigma for a long period after the first European intrusion. We do not know which islands still had the wild, indigenous Arawak vegetarians, which had been conquered by the fierce Carib cannibals, and which were empty. Most of the Arawaks and Caribs, as mentioned earlier, had been wiped out by the seventeenth century. In 1796 the British finally shipped all the Caribs they could find in their possessions to Honduras and Nicaragua, which may explain something of the turbulent history of those two countries. The Jamaicans, however, claim in their most romantic moods that some Caribs, either 'red' (pure) or 'black' (half-negro) were still on the island when the British arrived in 1655. However no Caribs were reported – only a total population of 3000, white and black – when Cromwell's troops occupied the island as part of his 'Grand Design' of defeating Spain by capturing her colonies.

The Spanish whites were made prisoners, while the blacks took to the hills and squatted. Their descendants, called maroons, squat in the hills to this day, and formed a focus of disaffection for the black slave population throughout the period of servitude. For fifteen years this great empty island, denuded of its productive population by circumstance, remained in a kind of suspended animation, acknowledged to be neither Spanish nor British. Port-Royal, near the modern Kingston, was the resort of outcasts and outlaws: buccaneers, thieves, convicts and whores. Their unrecorded life forms the basis of the tales of piracy, murder, treasure trove and seedy glamour which have thrilled or disgusted generations. They gave Jamaica a slow, semi-criminal start in the modern world.

Sugar and the slave trade developed in the last quarter of the seventeenth century, but it was not until the 1720s that Jamaican sugar production exceeded that of the much smaller Barbados. Jamaica, however, was a great entrepot for every trade in the West Indies, particularly the slave trade, and Kingston and Port-Royal were the natural centres for the whole of the British Caribbean. Yet the white population was tiny. It was not until after the American War of Independence that the white population of Jamaica exceeded that of Barbados. In 1783 Jamaica had about 20,000 whites compared to the 17,000 in Barbados; but while Barbados had declined in population to 160 whites per square mile at that date, huge Jamaica had only four. While in Barbados there were only four slaves per white, in Jamaica there were ten slaves to each white man, woman and child; not his property, perhaps, not his own personal servants, but with him on the island. And they were disaffected, even bitter, often driven to revolt.

Despite its emptiness, Jamaica had an aura of claustrophobia, induced perhaps by the mountains which form a spine along the island, with the peak called Blue Mountain, more than 7000 feet high, brooding and often cloud-capped, with forests all round, full of fierce maroons or other escaped slaves who were always ready to strike at the whites and to liberate their black brothers still in bondage. Jamaica never felt like a happy place.

Under slavery, Jamaica became an island of large, relatively unproductive estates. In 1783 there were over a hundred estates in all, each averaging more than 700 acres in extent, nearly four times the size of the average plantation in Barbados. An average estate would have more than 500 worker slaves, compared to under twenty on Barbados. Each worker slave in Jamaica produced only half the annual sugar crop of each slave in Barbados. Jamaica was therefore an island of low output per acre and low output per man, exacerbated by a difficult climate with a tendency to hurricanes and earthquakes. A condition of agricultural equilibrium was never achieved, as it was in Barbados. Because there was more land, it was not properly cultivated. Investment in slaves was far higher than that in land, perhaps four times as much. It is difficult to avoid the conclusion that when the economy of Jamaica became mature, in the period from 1770 to 1810, the interest of the trade became the engine for the survival of slavery itself. Through its sugar production Jamaica was the most important island, the one in which most capital was invested, the centre of the West Indies, and the entrepot of all trades involved in sugar; within twenty years it was to become the greatest sugar-exporter in the world.

By the beginning of the seventeenth century, every trader to West Africa was offered slaves in exchange for his European goods. Early traders sometimes refused to trade for slaves, and even in 1689, by which time what was known as the triangular trade had become established, many still had their doubts about the whole business. Most forcibly expressed were the sentiments of the philosopher John Locke at that date: 'Slavery is so vile and miserable a state of man, and so directly opposite to the generous temper and courage of our nation, that it is hardly possible than an Englishman, much less a gentleman, should plead for it.' Even if this were true, hundreds of Englishmen *became* 'gentlemen' through the slavery of others. In the seventeenth and eighteenth centuries the 'new men' of Bristol and Liverpool were elevating themselves in the social hierarchy of the country by means of the slave trade, and it is the phenomenon of the 'new men' which indicates not only why the triangular trade ever started, but also why it stopped when it did.

During the 1780s, the relative positions of England and any rival had been reversed. Despite the loss of the American colonies in 1783, the United Kingdom was strong enough to form the core of European opposition to Revolutionary, then Napoleonic, France; to fight a war lasting over twenty years, with one short intermission; and to win not only the war but a new Empire larger than that lost when the United States became independent. English capacity for recovery was obviously deep-seated, and led to a primacy in trade and commerce. This primacy was not suddenly discovered, nor was the dynamic which made England the first nation to enjoy the splendours and miseries of the Industrial Revolution. The causes go back a long way, and it may well be suggested that, whatever they were, they made England unique; not, let it be added, the United Kingdom, which included Scotland for most of the century and Ireland for none of it, but specifically England. Scotland and Ireland had their own problems, and made their own contribution to the history of the Americas, but a peculiarly English offering was made to the world by sugar.

The English had turned serfdom into a cash relationship between landlord and tenant far earlier than had most of the continental countries. The serfs might be pleased to become tenants, but it was the lords who really benefited. 'The English landed class turned lordship into ownership and instead of feudal lords became real owners of land. This was the most important change in the whole of English History.'[19] The money economy of English land tenure also made possible the institution of primogeniture in its most forceful manifestation. Without cash, settling a jointure (the income on a small amount of capital) on younger brothers and sisters is difficult to achieve. True primogeniture therefore must depend upon a cash economy in agriculture. Every English duke has worthy relations of an extremely humble nature, who live life as if they were in no way related; this is a social apartheid impossible to imagine existing in France or Iberia, England's rivals in the eighteenth century. Then, as now, some of these relatives were not so worthy, and in the past many of them sought their fortune on the high seas, in the colonies or in any trade which would yield a quicker profit than the traditional professions for the younger sons of the aristocracy – the army, the navy, the law and the church.

It was, and is, possible to engage in all sorts of pursuits abroad which would not be socially, morally or legally acceptable at home. One of the more disconcerting experiences for the upper-class Englishman is to find an erstwhile school chum engaged in some foreign country in an activity which at home would attract the attention of the law. Similarly, there is not an English peer of an old creation without cousins of various degrees of closeness engaged in every conceivable activity, legal or illegal. 'One draws the line', said an earl in 1983, 'at drugs. Otherwise my family appears to be involved in everything.' It could have been his ancestor, an ambassador, talking about his own collaterals two and a half centuries before – one of those collaterals was practically a pirate.

Besides the noble sprigs, there were the humbler and more numerous, if equally adventurous, relations of the landed gentry, or of the burgeoning merchant class, or unsettled younger sons, or illegitimate sons of professional men, or those who had supported the wrong side in any political dispute, including the Civil War. Finally there were the indentured, who might be petty criminals, or civil debtors, or merely those who had sold their labour for seven years in exchange for a free passage. The whole of society on the high seas, or in the colonies, was far more open than at home. ' . . . Broken traders, miserable debtors, penniless spendthrifts and discontented persons, travelling heads and scatterbrains. These and like humours first people the Indies and made them a kind of bedlam, for a short tyme. But from such brain-sick humours have come many solid and sober men, as these modern tymes testify.'[20]

After about 1680, one of the quickest ways to a fortune was in the triangular trade. This economically elegant development made the slave trade pay a dividend on every leg. In essence the trade was, as its name suggests, in three parts. The first leg was from England to West Africa, with trinkets and baubles (never gold), cast iron bars (the long bar, 9ft × 2ins × 4ins, later became a unit of barter), greycloth, firearms, gunpowder, shot, alcohol and salt. Apart from salt, none of these was necessary to anyone in Africa except to the native slave-trader who needed the guns for use and the other objects for trade. Significantly, salt is the only product on this list still exported in large quantities to West Africa from the UK. Native chiefs traded inland for captured slaves. Wars were often deliberately started for the sole purpose of capturing prisoners who were then sold as slaves. Sometimes both sides were in league with different white slave-traders. The whites found levels of price and quality variable and capricious, and barter might continue for days. Sooner or later, the ship set forth on the Middle Passage.

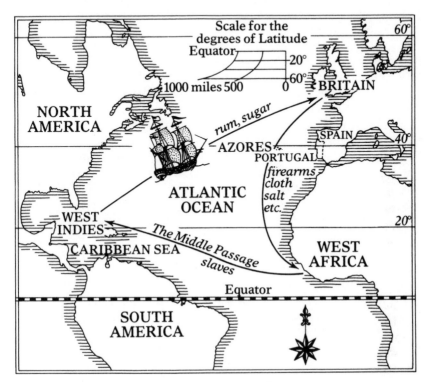

The Triangular Trade

This dreaded voyage was made unendurable if a ship was unlucky, or suffered from an incompetent crew so that it stagnated in the Doldrums for any appreciable time. The Doldrums roughly follow the real Equator, being north of 0° during the northern summer, and a ship becalmed in the Doldrums for more than a month in July–August might lose half its slaves and a quarter of its crew. In the northern winter the passage was cooler, faster and more profitable, so that ships tended to leave London, Bristol or Liverpool in the early autumn, plan to make the Middle Passage in December–February, and arrive in the Indies in the spring, returning to England again on the third leg, with a cargo of sugar and/or rum, in the milder northern summer.[21]

To prevent mutinies, and suicides by jumping overboard, male slaves were regimented, though women and children were sometimes allowed to run freely about the ship. There was a belief, common among the Ibos and Yoruba, but not confined to those tribes, that to jump overboard was a rapid way to return the soul to heaven. From this article of faith the slave, as a valuable piece of property, had to be protected. Adult males were chained together, each in a space about one-tenth of that available to a modern charter aircraft passenger, and there they lay prone, often in their own ordure, for up to three months. The stink, the imprisonment, the fear of the unknown, the inability to communicate, the strange white men, all these factors must have added to the natural horror of the sea voyage and help to explain why a number died on route,[22] however skilful the captain, however considerate and competent the crew, however easy the passage.

Having arrived in the Caribbean, the ship might go from island to island, but ultimately the slaves were taken ashore and sold. The vessel was converted from passenger ship to cargo carrier and loaded with rum, molasses and coarse, once-refined sugar for the rapid run home on the prevailing westerlies, a voyage of about thirty to fifty days. Allowing for storms, privateers, pirates, enemy ships, and timbers and rigging weakened by tropical pests, the ship would return home with or without a profit.

How handsome were the profits? For the survivors, they were high. In the early days, as many as a third of the ships might be lost to storm or human agency. During times of war, losses were higher.[23] But pirates were gradually brought under control by the Royal Navy. In the early days, the price fetched by a slave in the West Indies was as much as 700–800 per cent of the cost in West Africa (a buying price of £3 and a selling price of £25), but losses en route were very much higher, and the captains less skilled, the ships less suitable, the piracy worse and so forth. The trade became a vast industry, employing as many as several hundred British ships at any one time, and more of other nations. The price of slaves in West Africa rose as they had to be captured and marched from further and further inland, and so the profit margin was reduced to, say, the difference between £20 and £30, in other words a mark-up of 50 per cent. Loss of life en route dropped to a very creditable 5–10 per cent. The mature triangular trade was like any other trade, and the profits of all three legs had to be reckoned against the indebtedness of the planters, the price of sugar in Europe, the taxation and credit policies of British and foreign governments, and war and weather risks. The profits of the triangle followed profits in other trades and in other times. When risks were high, the successful made money and the failures were drowned, or faded from history in some other way. The mature trade was a stable institution become stodgy and dull, and returned only 25–35 per cent in the event of success, with all the downside risks still present.

It was the mercantilism of the triangular trade which really led to abolition, not the inhumanity. In 1790, the West Indian trade was easily the most important trading activity of the British nation:

	West Indies trade (including triangular)	East Indies trade (India, China, all Pacific and Indian Oceans)
Capital employed but not real property, slaves, etc. abroad	£70,000,000	£18,000,000
Exports from UK (excluding bullion)	£ 3,800,000	£ 1,500,000
Imports into UK	£ 7,600,000	£ 5,000,000
Duties paid to government in UK	£ 1,800,000	£ 800,000
Tonnage employed	300,000 tons	160,000 tons

The figures in all cases exclude government transports, troop movements, stores, and of course naval vessels.[24]

The West Indian merchants carried such weight that the sugar islands were considered in 1763 as being more important than Canada, both British and French, and by one section of politicians as being more vital even than the American colonies. The West Indian interest was always unpopular because of its arrogance, its get-rich-quick propensity, its attractions for nature's gamblers, and its faint taint of slavery, which carried into the most elegant drawing room – slavery clung to the West Indian grandee as does horse-dung to the racehorse-owner. The opposition to sugar and slavery was not confined to the new romantics of the late eighteenth century, to the dissenters, or to the staunchest and earliest of abolitionists, the Quakers. Dr Johnson described Jamaica as 'a place of great wealth, a den of tyrants, and a dungeon of slaves', and proposed a toast at Oxford to 'success to the next revolt of the negroes in the West Indies'. Earlier, Alexander Pope had quoted with approval the lines from the *Odyssey*: 'Jove fixed it certain, that, whatever day/Made man a slave took half his worth away'. From a more calculating, economic viewpoint, Adam Smith and Jeremy Bentham both endorsed Jove's calculation of a 50 per cent loss of productivity in the process of enslaving blacks. Thoughtful men had got out of both slavery and the slave trade early in the eighteenth century. The early fortunes made in Bristol gave way to a complicated, involved trade based in Liverpool and founded on lines of credit, Manchester manufacturers, Birmingham gunsmiths, gunpowder from Cheshire, and ironmasters from all over. The trade was vertical. Sugar was said to be supporting 'half Lancashire and a quarter of all British shipping'. Yet in mercantilist terms, the triangular trade in its maturity made no sense at all.[25]

In the decade 1783–93 the British controlled, by one means or another, more than half of all the sugar trade between the New World and the Old. Yet, of that near 60 per cent, Britain herself consumed almost enough to say that half the sugar brought across the Atlantic went down British throats and out to sea via the drains. In other words, perhaps 25 per cent of English maritime effort, the work of a quarter of a million workers in England, and the sunbaked effort of all those whites and slaves in the colonies – all this was flushed out of the consumer within hours of being consumed.

Each year during this decade the United Kingdom ingested about 70,000 tons of refined sugar, which replaced about 80,000 tons of wheat in energy value. The wheat was worth during that decade an average of £10.5 per ton. Wheat was roughly in balance during the ten harvests 1783–92, and the UK imported during that period an average of less than 15,000 tons, or about 0.5 per cent of production, of which most came from Ireland. In most years an extra 80,000 tons of wheat would have been available from the Baltic, from the Americas and from Ireland. The entire sugar effort, costing the British about £5–6 million net per annum, replaced wheat which would have cost not much more than £800,000 net per annum. Such is the price, not just of addiction, but of corruption.

The corruption was not only of the consumers, poor addicts that they were; nor of the slaves, degraded by the system which needed them; nor of the slave-owners, objects of both hatred and contempt; but of the whole mercantile system as it existed at the end of the American Revolution. Whig mercantilism ('parliamentary Colbertism') became suspect to the generation which made peace with the Americans: indeed had not the Whig policy led to the loss of America? The Elder Pitt had said so: was this not enough for his son, and his son's friends and allies? The Younger Pitt came to power convinced that Tory principles should prevail. Trade was to benefit the consumer, not to enrich the native industrialist. Imports were to increase comfort, even luxury, among the people of the importing nation: it was no longer seen as an evil to import in the absence of home production. Exports were necessary, according to the new principles as enunciated by Adam Smith and grasped by Pitt, not to gain and hoard bullion, but to benefit the individuals who made, traded and transported the exports. The sum of the happiness of individuals became the greatest good. The sum of the individuals' gain became the wealth of the nation.

Similar sentiments were voiced all over the civilized world, particularly in the new United States, and in France, where the Encyclopaedists had defeated the ideas of Colbert. During the decade 1783–93 the foundation of the free trade world was laid by Pitt and his friends. But for the French Revolution and the subsequent wars the prosperity of that decade might have continued for a further generation, to benefit the whole world instead of only America and England. It was this opportunity which gave both countries a head start in the next century.

As far as the triangular trade was concerned, the new men hated the closeted, fetid corruption of the City of London and the West Indian interest. To this hatred of mercantilism, which gave the early free traders a moral edge which they never entirely lost, was added the philanthropists' loathing for the callousness of the slave trade and, worse perhaps, its mercantile justification. If the mercantile, economic justification for 'the system' could be blown away by the fresh air of individual responsibility and the individual contribution to the wealth of nations, would not the moral base for slavery disappear?

Thus was born the great alliance of the Tories, under the Younger Pitt, and the philanthropists, under his friend William Wilberforce. Mercantilism was dead as an intellectual belief with any respectability. 'Parliamentary Colbertism' was as discredited as the slave trade became.

The new men believed in free trade, and it was a natural alliance to join with those who believed in free men. Both parties saw an immense amount of activity – and misery – which benefited only those who were unworthy, since their profits rested on an illegitimate slave trade. For both practical and moral reasons, it was agreed among the abolitionists that the slave trade should be the first target. It was more horrible, more susceptible to easy abolition, and it would not, above all, need any parliamentary money for compensation. Even so, it took time. Five times Wilberforce was defeated in the Commons, but in the end he triumphed. Within eighteen months of Trafalgar the slave trade would meet its own defeat, and thereafter be considered as foreign and as unworthy as any other enemy.

*

Battles are gained or lost because the winner is stronger or the loser is weaker; that is obvious. But in the fight between reform and resistance it is not the strength of reform, but the virtue of reason and circumstance, which overcomes resistance. Reform is no boxing match, nor a football game, nor a contest between past and future, nor a battle between good and evil. It is the importance of circumstance which makes reform possible, and allows reformers to claim that their point of view has the inevitability of logic. So it was with the abolition of the slave trade. Very few could defend the system once four circumstances obtained.

The first was that clever money and 'new men' were moving elsewhere, into England's own Industrial Revolution, into turnpikes and canals, and into trade outside the triangle. Above all, the bankers were beginning to see that the sugar trade was an incubus in their balance sheets. The planters and traders suffered from hard-core borrowing, on which the debtors were hard-pressed to pay interest, let alone repay the principal. The triangular trade was a mature, staple trade, and clever money does not remain long in mature trades.

Secondly, because Britain controlled the seas after Trafalgar, no sugar or slaves could be moved except by permission of the Royal Navy. Rivals could not therefore cheat by engaging in the slave trade or importing sugar produced by cheap slaves without the Royal Navy's permission, and when peace came the slave trade, already 'abolished' by the United States in 1794 and by Denmark in 1803, could be universally condemned by *force majeure*,[26] so that no other country could steal British markets with cheaper, slave sugar. Or *was* slave sugar cheaper?

This was the third reason. No one knew whether sugar was produced more cheaply by slaves or by free blacks. But bankers, many of whom were Quakers, suspected that it could be cheaper to use free labour. What was quite clear was that, if the slave *trade* was stopped, the value of the remaining slaves in the Indies would rise. Abolition would thus lead not only to a rise in capital values, but also, inexorably, to better treatment for adult slaves, because they could not be replaced in the local market. Thus were Christian morality and banking prudence combined to deal with an institution viewed with distate, yet representing an enormous investment.

The fourth reason which led reform to overcome resistance was war itself. The triangular trade could not be abolished in peacetime without compensation. But in the wartime circumstances of 1807, the nation could use every ton of shipping Britain had, and freight rates were very high. So all parties were compensated without cost. The planters had a rising value in their slaves, the bankers had safer loans, and the shipping interests had immediate other employment for their ships.

It used to be a Victorian morality tale that good men such as Clarkson and Wilberforce triumphed over the evil of slavery, and it was a post-Freudian piece of sourpuss logic that Wilberforce & Co. were more interested in their own salvation than they were in the poor slaves. But an examination of the writings of the time does not support either of these views.

The *Encyclopedia Britannica* of 1792, the third edition, was a progressive publication edited in Edinburgh. The authors approved of the American Revolution, and of the course of the French Revolution to date – Louis XVI was not guillotined until after the *Encyclopedia* went to press. The slave trade was condemned as an immoral cruelty which was economically inefficient. Yet the negro was described thus:

> Vices the most notorious seem to be the portion of this unhappy race: idleness; treachery; revenge; cruelty; impudence; stealing; lying; profanity; debauchery; nastiness and intemperance all said to have extinguished the principles of natural law, and to have silenced the reproofs of conscience. They are strangers to every sentiment of compassion, and are an awful example of the corruption of man when left to himself.

Progressive opinion at the end of the eighteenth century could not therefore be called unprejudiced. Nor was slavery condemned in itself. It was the *trade* which outraged progressives of the day, and it is impossible to avoid the conclusion that at any propitious moment the logic of the situation was that the trade, as opposed to the institution, would be abolished. And this is what happened, when all the conditions involving the planters, the bankers and the shipping industry were favourable.

From 1807 onwards the sugar industry of the British West Indies went into slow, then precipitate, decline, Jamaican production falling from 100,000 tons in 1801 to under 5000 tons in 1913. The decline from a wartime peak in 1801–5 preceded the abolition of slavery itself by a generation. During the Napoleonic Wars, a specifically European development would prevent the return of the Caribbean sugar trade to any kind of boom.

Sugar was the first food (or drug), dependence upon which led Europeans to establish tropical monoculture to satisfy their own addiction. Because sugar cane was a labour-intensive crop, the ratio of slaves/sugar always remained at least ten times greater than the ratios of slaves/tobacco, or slaves/cotton, or any other crop grown in servitude. Perhaps three-quarters of all the Africans transported across the Atlantic, possibly as many as 15 million out of a total of 20 million enslaved in Africa, must be debited to sugar. Yet after 1750 the husbandry existed to grow a different form of sugar in western Europe itself. Only the motivation was absent.

During the Napoleonic Wars French sugar ships had to run the blockade of the Royal Navy in the West Indies. They also suffered the loss of 100,000 tons of sugar a year which had previously come from Dominica (Haiti). Against this background of shortage and high prices Napoleon became aware of the botanical researches of Andrew Sigismond Margraf, of the Berlin Academy. Margraf had discovered that there were significant quantities of sugar in carrots, parsnips and, above all, sea beet, a cousin of the beetroot and the mangold. Any child knew that ripe roots were sweet, but it was Margraf who first isolated the sugar in roots. However it was not until 1801, long after Margraf's death, that any practical work took place. Encouraged by the high wartime price of sugar, the first selection and crossing of roots for sugar took place. Within ten years the beet industry was born, and in many continental countries, notably France, it was developed with much dirigisme[27] and subsidy after the peace in 1815. The French did not wish ever again to be dependent for their sugar on the goodwill of the Royal Navy.

Within fifteen years, beet sugar was threatening the tropical trade. Within thirty years, cane sugar had lost most of its continental markets. In 1885 beet overtook cane in the total world trade in sugar. In some poor, primitive areas of eastern Europe, the only sugar which has ever appeared as an ordinary grocery staple was made from beet. In the late twentieth century every European country can be self-sufficient in (beet) sugar if it so wishes.

Against this background, the abolition of the British slave trade in 1807 was followed in 1834 by the emancipation of the British slaves. The reason for the delay in bringing about emancipation was the problem of the formula by which the slaves could become ordinary workers, and the question of compensation. These were intimately linked. If the slaves became land-owning peasants there would be no one to work the estates, and the land would become worthless, as in the French colonies. If the slaves became wage-earners, they would not know what to do with the money because their basic needs had always been supplied for them. In the end, it was decided that the slaves were to become apprentices for seven years, and the owners were compensated by up to 60 per cent of the value of the slaves, on the grounds that the apprenticeship period was equivalent to the missing 40 per cent money value. Everyone was happy except the planters. The bankers saw their money again – nearly all the emancipation money went straight back to the City of London to repay the planters' loans, and was probably reinvested in railways at home. The dissenters and the liberal middle casses assuaged their consciences by the use of government money to right a wrong, just like any modern socialist. The slaves got their freedom. Only the planters had a problem: instead of working, the 'apprentices' ran off to find a piece of ground to live on, or squatted in some corner of a sugar field and raised yams, plantains and bananas, and thumbed their noses at the overseer. Confidence fell with sugar production, which halved in five years. It was to halve again, and again and again.

Ten years after emancipation, when sugar production in the British West Indies was already in steep decline, the free trade party in England carried the day, and all quotas and differential duties were removed on a graduated scale. Parity between British and foreign sugar was to be achieved by 1851. The West Indian interest was effectively bankrupt, if not silent. Between 1832 and 1848 there were 105 trading failures, amounting to bankruptcy, among the merchants in England. There were over a thousand bankruptcies in the West Indies in the fifteen years after emancipation. The planters blamed emancipation itself. But they were wrong. They should have blamed the bankers, and all the other supporters of free trade.

The wise bankers never lent any more money to the planters. The clever bankers had got out of any commitment to the West Indies years before, during the Napoleonic Wars when prices were high. The wisest of all had dropped support of the West Indian interest during the earlier booms, notably during the Seven Years' War of 1756–63 or during the American War of Independence, 1776–83. The planters never learnt that the virtue of clever money is not the money, which is the same as anyone else's, but the free advice which goes with it.

Under-capitalized, without any workforce disciplined by the need to earn money, their land valueless without labour, their former lifestyle rotting in the tropical depression, the planters in the British West Indies were never to recover. Sugar production fell until World War I, when the demand for sugar could no longer be satisfied by the usual beet imports from Europe, a situation which temporarily boosted the West Indian sugar cane industry, in particular that of Jamaica. But the great days were over. The sugar addiction was to grow, but it would be satisfied more cheaply by beet, and by the new methods and better land of Cuba, the quintessential nineteenth-century sugar island.

Cuba had been discovered and claimed for the King of Spain by Columbus in 1492. He returned to Spain with tobacco and syphilis as the first imports into Europe from the Western Hemisphere. For nearly three centuries this huge island, nearly as big as England, was populated by less than 200,000 people, white and black, and all the permutations possible between the two races. Havana, the capital, less than 200 miles from mainland Florida, accounted for more than a third of the population. Cuba was ruled until 1820 by the Spanish viceroy in Mexico and, after that date, formed with

Puerto Rico Spain's only remaining colonies in her once huge American empire. Non-commercial colonists had tried to impose on the land of Cuba the plants of Spain: vines, olives and wheat. None of these prospered, but the cattle brought over in the sixteenth century multiplied exceedingly in the hills, so that dried beef was exported to Mexico and Venezuela, and provisioned ships.[28] Leather was exported to Spain, and remained Cuba's most important export until the end of the eighteenth century.[29]

For the Spanish, silver and gold was an annual treasure tribute from the Americas, and the immensely valuable treasure fleet lay in Havana harbour, ignoring the possibilities of the island which was to become the greatest sugar-producer in the world. European Spain, like Portugal and Sicily, was self-sufficient in sugar, and trade between Spanish colonies was only permitted with Spain herself – and for three centuries only with Seville. The fleet left Havana for Spain fully provisioned, provided with citrus fruits against scurvy, and green vegetables as well as dried beef. These commodities came from plants or seeds transferred from Europe, probably via the Canary Islands, which also provided a large number of the poorer white immigrants. The city of Havana, bigger than any eighteenth-century English city except London, was a whore among towns, existing only for the fleet, providing the waiting seamen with what seamen have always thought they wanted: drink, gambling and women. In 1762, when the English captured and held western Cuba for less than a year, Havana had already acquired its raffish, *demi-mondaine* reputation, which was only brought to an end by the bourgeois virtues of Communism two centuries later, in the 1960s.

It is important to remember that in the Americas the Spanish instinctively reproduced the conditions of sixteenth-century Spain: large cities with cathedrals, civil and military governors, market towns and villages into which were gathered the rural population. The Spanish were a much more urban people than the English in the Caribbean. The English islands, correspondingly, were typically covered with estates of various sizes, as at home.

In the short time available to the English in 1762–3 they could not change much of Cuban trade, but they did open up men's minds to the possibilities of sugar-growing. Cuba was already an important tobacco-producer (snuff, not cigars: there were more than a hundred snuff mills in 1763, and more than two hundred by the end of the century), and from the Indians, long since expired, the Spanish had learnt how to grow sweet potatoes, yams, bananas, maize, most of the beans we now know, yuccas and the American pumpkins and squashes. The Spaniards were not as efficient as the Indians, and were wasteful of land – as indeed they could afford to be. In 1763 the density per square mile was less than five, far less than in medieval Spain or Italy. A hundred years later it was eight times as much, with a total population of 1.6 million and a density of 40 per square mile. It was still not very crowded, but the difference was caused by sugar.

The introduction of sugar culture entailed not only the discovery of markets other than Spain – being self-sufficient in sugar – but also the entry of Cuba into world trade, including the slave trade. While there might have been a few thousand black slaves introduced into Cuba before sugar, they were relatively well treated and reproduced themselves without difficulty in an island devoid of the intense pressures apparently inevitable in commercial sugar production. After the 1770s, possibly as many as 2 million may have been imported. The very last recorded import was of 600 slaves in October 1865, on the estate of Don Marty: 'The landing is denied by the authorities, but the fact was publicly known at the time.'[30]

Cuba's indented coastline made smuggling very much easier than it had been in the British islands, and thousands of slaves were landed during the century before 1865, directly on sugar estates and not through ports. This makes statistical analysis difficult, but it is quite clear that at least 500,000, and perhaps more, were landed between 1830 and 1865. The death rate on the plantations averaged at least 10 per cent a year, no better than a century before in the French or British colonies.

To supplement these illegal negro imports, between 1847 and 1880 about 140,000 Chinese coolies were shipped round the Horn, or via the Cape of Good Hope, a 15,000-mile voyage. Tougher than negroes and less fatalistic, only 11–12 per cent died en route. Less than 25 per cent survived in Cuba, however, and less than 1 per cent returned to China – because they only cost the passage money, these 'indentured labourers' were in many ways even less well treated than slaves.

The conditions of the slave trade itself, from Africa to the Americas, did not improve after it had been made internationally illegal.[31] On the contrary, for obvious reasons it was probably far less humane. The moral must be that, to be effective, the reformer has to guarantee that the abuse abolished does not leave behind an illegal operation more heinous than the open one. The Americans in particular, who defied the ban on the trade, exhibited the same contempt for the new morality as their spiritual descendants showed for Prohibition a century later. Laws were made not to be obeyed, but to be circumnavigated.

After 1820 the Royal Navy maintained continuous patrols off the Bight of Benin in West Africa, in the trade wind routes, and of course throughout the Caribbean. They were an excellent means of keeping peacetime crews alert and well-exercised, and constituted an elegant way of justifying naval expenditure to those who were the natural 'economizers' in the nation, and who did not really believe in the 'Big Navy'. The ships were often commanded and crewed by men who had seen enough of the horrors of the new slave trade to approach their task with an evangelical zeal.

Zeal was needed. Losses at sea were now higher, and ships were faster. Slavers were prizes, even in peacetime, so that every man's hand would be against them, and the slave ships therefore avoided contact with other vessels by using tortuous routes. Disease was rampant. One slave ship was found by the Royal Navy, floating inert, its entire complement, black and white, slave and crew, blinded with opthalmia, and only able to grope about the vessel. Not unnaturally, they were starving in the midst of plenty.

Cuba was, like the ocean, an invitation to avoid the law. The law was such an ass, for a start. The internal slave trade between Cuban provinces was declared illegal from 1820, but was ignored in the same way as the same law in the United States. Trade with any country other than Spain was illegal until the 1820s, and *that* law had been ignored for at least two hundred years. Merchants were only allowed to use Spain's own creaking banking system, and all letters of credit or other instruments of payment had to be drawn up more than 3000 miles away. As a result, a great many quite large transactions were settled in coin. Most of the external trade of Cuba was in the hands of the British and Americans of all kinds, both from the USA and from Latin America.

It was the English who had introduced sugar into the island, in their short tenure during 1762–3. It was the Americans, however, who recognized the virtues of the fertile land after the emancipation of the slaves in the Southern States in 1865.

Slave-sugar had been produced in Louisiana and other Gulf states before the Civil War, and had proved to be very profitable. But after the abolition of slavery, Southern men and Northern capital abandoned the cane-brakes of the Deep South in favour of the new opportunities in Cuba. Sugar production in the United States had proved to be as destructive of the slaves as in any other servile economy, and 'sugar-slavery' became a hated, much feared threat for thousands of cotton slaves who knew nothing of the cane-brakes. Life expectation in sugar slavery even in the year before the Civil War was only half that in other forms of field slavery, and the prospect of working in the sugar estates, even as free men, was not attractive to the blacks of the Deep South. Americans therefore transferred their attention to Cuba.

The cultivation of sugar cane requires, as has been noted, good land, fuel for the refinery, ample labour, a home market which can supply the wants of the colony, machetes for cutting cane, cauldrons for boiling sugar, and clothing for masters and slaves. In Cuba, as in all the Caribbean colonies in the late eighteenth century, only land and fuel were freely available, and the shortage of labour was made acute by disease. Although by the late eighteenth century malaria was contained by quinine, yellow fever was still rampant and mosquito-spread, and not yet known to be a disease associated with dirt. It was endemic in Cuba until the Americans eradicated the dirt and thus the disease in the early twentieth century.

Slaves provided the labour in the sugar plantations, and whites the artisans. Cuba had hardly any Indians left, if any, in the eighteenth century. Spaniards from Europe, known as *peninsulares*, did not stay long, and usually went back to Spain after a tour of duty. If they did stay, and brought up a family in Cuba, the children became known as *criollos*, or in English, creoles; in English the word has mulatto (half-caste) connotations, but not in Spanish. There were also half-castes, quadroons and mutations.[32] Spaniards were far less racialist than the English. There were few taboos against intermarriage in the colonies, and the taboos were more of class than of race. Large numbers of Spaniards in Spain itself have some sort of black ancestry, as in other Mediterranean countries and Portugal.

As slaves became 'illegal' and therefore more expensive in the early 1820s, machinery was developed to save labour. Crushing rollers and copper boilers were integrated and steam-driven, using for fuel the cane residue known as bagasse. The oxen which had formerly supplied the power were no longer needed, nor were the slaves who had tended them.

After 1855 another great improvement took place: a central sugar mill, if necessary miles from the plantations, would service several thousand acres connected by a railway which was often owned by a private American company.[33] The first Cuban railway – indeed the first in Latin America – was completed in 1845. It ran between Havana and Guines, a distance of 45 miles.

While all these labour-saving devices were employed, using American, French and British capital, whites of the technician or foreman class, from Europe or the Canaries, would manage the machinery, and contract Europeans would oversee the labour, including slaves and Chinese coolies. Technical improvement was continuous, and the industry survived revolts and civil disturbances. Sugar production per slave per year increased from 200 lb in 1760 to about 1 ton in 1840 and about 3 tons in 1880, the year slavery was finally abolished. The whole of this advance was the result of mechanization of transport and refining, and there was very little increase of productivity in the field. By the end of the nineteenth century Cuba was producing ten times as much sugar as Jamaica at the beginning of it, with only three times the labour and about five times as much investment in machinery as in land. Of the million tons produced in the 1890s, more than three-quarters came from US interests. The rapid, forced growth of sugar production was to increase again by a factor of five by 1925, when production reached 5 million tons.[34]

This huge, fertile island never seemed able to live in peace with itself during the second half of the nineteenth century. Four American Presidents, two before the Civil War and two afterwards, wanted to buy and annex Cuba. Some of the Spanish *peninsulares* wanted to join with the *criollos* to demand independence from Spain. Others wanted to unite with Mexico; yet others to enjoy a semi-colonial status with America; and others to become an American state. Before the Cuban War of Independence of 1895–1900 there was never a majority of whites, let alone mulattoes and blacks, for any single policy.

The situation erupted and even involved the young Winston Churchill, who as a war correspondent saw some incompetent military action on the Spanish side. The US battleship *Maine* blew up, with great loss of life, in Havana harbour in February 1898. The explosion was in fact caused by stale, unstable cordite, though everyone thought at the time that the explosion had been caused by an external Spanish mine. War was declared by the United States in April 1898, and was over within months. The Americans brought an end to the endemic yellow fever as well as to three hundred years of Spanish rule. American intervention guaranteed independence from Spain. It also guaranteed that American capital would remain in some kind of dominant role throughout the first half of the twentieth century; this was succeeded by the 'Good Neighbour' policy of the 1940s.[35]

The Caribbean ('America's backyard pond') has enjoyed a past quite different from almost all of the United States, Louisiana and New Orleans perhaps excepted. Cuba, despite the excellence of its tobacco and the attempt in the early nineteenth century to grow coffee, has been dominated by sugar for nearly two centuries. The island was inevitably neglected by the Spaniards during the Napoleonic Wars and plantations were allowed to develop on laissez-faire rather than mercantilistic lines. Slaves were the losers, but sugar became relatively cheaper, and the cheapening of sugar, and its disposal, became a paramount consideration.

Sugar had needed a huge labour force in the days of slavery. In Cuba, for the first time, 'free' workers plus machinery plus good management showed how the whole growing process could be made cheaper, and sugar came down in price to a point very much nearer to that of the alternative sources of energy. Then the markets became the problem. In good times the world is capable of producing four to five times as much sugar as the most addicted population could ever need. Cuba showed the way, and also revealed the vital post-1945 nature of the market. The supply side problem had been solved; it was demand that was deficient. The United States guaranteed, effectively, to buy every ton of sugar that Cuba could produce during World War II and during the postwar shortage. Cuba responded. But the USA did not continue to purchase every ton once the worldwide shortage was over in the early 1950s. It was, in a sense, *Yanqui*[36] inability to absorb everything that Cuba might produce which in turn caused the economic stagnation in Cuba which made Fidel Castro possible. But Cuba, like many other places, could not meet the social challenge of successful primary production, which inevitably reduces employment as it cheapens output. While it had become the world's largest-ever sugar-exporter, Cuba had never developed political institutions suitable for a twentieth-century state. Communism beckoned with its meretricious, counterfeit solution.

Fidel Castro is the child of an ex-soldier from Galicia in Spain. His father was a tough, sometimes violent man who made himself the sugar-rich landlord of 10,000 acres. His mother was once the cook in his father's mansion. On to this genetic omelette Castro sprinkled a few years of study, more of revolutionary effort, and still more of government. On the way, he and his allies brought us to the brink of World War III in 1962; attempted to communize Latin America in the 1960s; and tried to help the USSR mobilize the new nations of Africa against the West in the 1970s. Cuba continues to destabilize and impoverish many areas of the world.

Meanwhile, there is little evidence that Cuba has done more than substitute one sugar buyer for another. Even the tonnages, though now a state secret, seem to be about the same. Of all the ex-sugar dependencies, Cuba, which had so much opportunity during independence, is perhaps the saddest failure. It is now only another sort of colony, feeding the USSR with the sugar its inefficient beet industry cannot provide, and which its people crave. Though Castro has tried to reduce his dependence upon sugar, the addiction to produce remains as strong as ever, and Cuba's colonial status is confirmed. It is nearly five hundred years since Colombus discovered it, and Cuba has had only three mother countries in all: Spain, for four centuries; America, for rather more than fifty years; and the USSR, for rather less than thirty. The quincentenary of Columbus, and of Cuba, falls in 1992. It should prove interesting.

What has sugar done for the world? We all know what it has done for the consumer, and there is a vast literature on the subject of rotten teeth, of impaired digestive systems, and of addicted nervous dependence. Sugar is such an easy substance to consume and so cheap in the twentieth century that the food-manufacturing industry has put some kind of sugar into almost every food, from bread through beans and soup to sausages and convenience meat dishes. Because they eat sugar people are fatter, more constipated, vitamin-deficient, more prone to disease, inclined to alcoholism and unable to avoid the dentist. But the effect on the Caribbean has been even worse.

The Amerindians have disappeared almost without trace. Not even in the United States or in Australia is the original, indigenous population so conspicuous by its absence. The white elite left the West Indies long ago, unless it is here once again (for a few years) to make a new fortune out of tourism, minerals or bananas. Few islands in the Caribbean have ever made any concerted effort at self-sufficiency, and, despite the fact that the Caribbean has more indigenous food plants than Europe, people even in some of the more favoured agricultural areas would starve but for imports, usually from Canada or the United States. A permanent trade deficit exists between this region and the rest of the world. There is an absence of know-how, a disrespect for the connection between hard work and profit, between the long haul and success, and between problems, incentives and solutions. Who can blame the inhabitants? They have faith in flash remedies for social and political despair, and a tendency to favour the less rational forms of religion, including various forms of Black Islam, voodoo mixed with Catholicism, spiritualism mixed with Protestant sects of various kinds, and animism mixed with God knows what.

Though bad enough, these are not the worst effects of sugar. The people of the Caribbean are quite different from the modern African. They have to carry on their backs a couple of hundred years of slavery, on average, but that does not explain the whole difference. The inhabitants are descended from the survivors of slaves who were in the same place for no reason of community. The crucial point is that many islands are over-crowded with incompatible people far beyond any ability of their adopted homeland to support them in dignity. The culture is raw, new, imperma-nent, ugly and insubstantial, and gives grave cause for concern to all who love the Caribbean, which is increasingly unable to resist further encroachment from outside.

Perhaps 20 million, perhaps more, Africans were torn from their homes and the survivors transported to the Western Hemisphere, so that today there is more negro blood in the Caribbean than in Africa; the ancestors of three-quarters of these displaced people were brought over to meet the demands of the white man's sweet tooth. This displacement did not stop in the West Indies. Probably half the slaves in the USA originally came through the Caribbean.

In modern times, since World War II, West Indians have emigrated from their rural slums to their own cities and in huge numbers to Britain and the United States. The language, the music, the traditions of slaves have transformed parts of American and European popular culture.

Attitudes formed during the 440 years of the slave trade make integration difficult, if not impossible. The multi-racial society is a mirage, always somewhere ahead, never here today. The Caribbean contains some of the most beautiful islands in the world, but it is difficult to enjoy the present when the brooding past is remembered.

These islands had one great commercial asset in the days of slavery – the work done by slaves. So much were the slaves the most valued commodity that in the wars of the eighteenth century, or the raids which occurred between the wars, they were the prime target. They were removed or killed in order to damage the enemy. This reinforced the lack of community feeling between the individual islands, which were owned by different foreign powers.

Abolition of the slave trade, followed by peace, improved the slaves' lives, but when emancipation came the hollowness of the lifestyle was exposed. The British islands were without schools, or roads, or sewage or water supplies. None of these was considered 'necessary' for agricultural slaves: parts of the Old Country also lacked amenities, but there was a thriving middle class which would cry out for these advances in every town in England during the nineteenth century. There was no middle class in the islands. There were no poor whites, nor middling whites, nor actively mobile whites of any kind, as in the American South. There were the former owners, if still there, and the former slaves, who had nowhere else to go.

What parallel can produce an analogy? There is none, but the nearest might be the abandoned mining towns of America. In overcrowded Europe, India and China there are few such places, but they exist in the Appalachians, the Rockies and the Sierra Nevada in California. To understand what an English Caribbean island was like before the tourist boom, imagine an old western ghost town, with the doors swinging in the wind at the entrance of the saloon. No one is in town except for a dotty old man who wheezes his memories in exchange for a few pennies. In the Caribbean there was, at the same time, an increasing population of idle people, without any function except subsistence, unproductive, outside the world economy.

The shiny new resorts have brought some of the population back into that economy, but there are too many people in the wrong places, and it will take generations to build the kind of sense of community which the humblest African peasant takes for granted.

Notes

1 Before examining this proposition, it is worth noting that the selling ploy conveniently ignored the fact that East India sugar estates used labour whose bondage was as servile as that of the negro in the Caribbean. But the English buyer of sugar was not always to know of this aberration.

2 Sugar cane is a tall, jointed reed, 3–9 feet high when 'mature' each year. Canes need deep, rich soil, and in the right place plants will survive for eight to ten years. In less suitable areas, pieces of cane must be replanted more frequently. Sugar cane does not survive frost, and temperatures just above freezing are just as damaging to it. Cane is harvested by cutting the stalks at the right point, a task which demands skill as well as strength.

3 'Creole' is the anglicized form of the French word of the same spelling, derived in turn from the Spanish *criollo*, which is a negro diminutive of *criadillo*, meaning 'born, bred, domesticated'. 'Creole' means naturalized to and born in the Caribbean (New Orleans and Venezuela as well as the islands), and is a term applied to people, animals and plants. Alexander von Humboldt, the great German naturalist, noted in 1800 three different forms of sugar cane in the Caribbean: the Creole, the Otaheite, which had been brought from the Sandwich Islands (Hawaii), and the Batavian, which came from what is now Indonesia. 'Creole' as a word had no racist meanings when first used: a creole white was one born in the Caribbean, while a creole negro was one born there and not in Africa. It is a mistake to imagine that the word implies a degree of cross-breeding, whether of humans, animals or plants.

4 Unextracted honey – honey still mixed with beeswax – has had connections with purification in India, China and Egypt. After the Exodus and in the desert, the Jews were forbidden to use meat, honey or leavened bread in any form of sacrifice (Leviticus 2:11). This may have been because honey was used by the Egyptians as a purity symbol, or more probably because the 'land of milk and honey' towards which they were moving supported so many bees that honey was a mundane product, never to be considered holy. Honey and wax were regarded as holy in Greece and Etruria, and to many tribes in Africa. Beeswax for Christian churches originally came from the best 'virgin' honey, that is from a young colony of bees which have never swarmed. Was the process of reproduction in Christian eyes a dubious proposition, even when conducted by bees?

5 The date at which the Spanish started growing sugar cane in large quantity; also the time when the Venetians took over Cyprus, a major sugar-producing country.

6 It may seemed surprising that an expensive luxury such as sugar should become popular so quickly when honey was freely and widely available. There were three main reasons for the rejection of honey in favour of sugar. Even when very pure indeed, honey always imparts a taste which may be inappropriate to the accompanying food or drink. Before the nineteenth century, in any case, pure honey was almost impossible to achieve: the extraction of honey from the comb was a laborious and difficult task, and the honey always included a certain amount of wax, which imparted a taste that was pleasant enough, but stronger than that of the honey itself. Honey also frequently contains substances to which many individuals are allergic.

7 See pp. vii–viii.

8 Slavery has always possessed the same inexorable logic as any other form of economic organization. Few slaves = humane treatment. Many slaves = cheap lives = inhuman treatment and fear. Whenever a high proportion of the population consisted of slaves this drove their masters to unconscious cruelty and habitual kennelling of their slaves away from all free men, often permanently in chains. Slaves frequently spoke only a foreign language, and were strangers from capture to death. The revolt of Spartacus (73–70 BC), which shook the Roman Republic to its foundations, made the relationship between masters and slaves much more severe; so did the sale of huge numbers of slaves by triumphant generals such as Crassus and Julius Caesar. At one time during the reign of Augustus the slave population may have outnumbered that of free persons. This kind of ratio leads to all the worst effects, in master and slave alike.

9 Feudalism is the relationship between vassal (tenant) and lord (landlord) which is non-mercenary. 'Rent' originally took the form of service rather than money.

10 This problem of slavery, serfdom and free workers is also one of relationship between master and man. For the 'noble' an air of hereditary superiority was an essential from early childhood. For superiority to have validity it is necessary that the dependents should be made to feel inferior at all times. It is to this factor, as necessary in feudalism as in slave societies, that we may ascribe white male attitudes to gender and race. The inferiors also deferentially accepted the situation, and sought to make the best of a bad job. At all times, it is necessary to remember that for all the Mediterranean peoples, and most Europeans, until at least AD 1500 the next world was more important than this. For today's incompetent, disabled or unfortunate, the compensation of heaven is usually no longer a credible option.

11 A monocultural system involves growing the same crop on the same land, continuously.

12 A reference to the prophecy in Genesis 9:25–6 that the children of Ham should be servants to their brethren; one of the 'good Christian's' justifications for the slave trade. Ham is held to be the Egyptian word *khem*, meaning black. In Psalms 78:51, 105:23 and 27, and 106:22 Ham denoted Egypt. No one knows whether the 'blackness' of Egypt referred to the soil of the Nile valley, or the people of those times.

13 The Emperor was a humane man. He had already 'emancipated' the slaves in 1542, but this was disregarded by his subjects, as was the Bull of Pope Leo X in 1514: 'Not only the Christian religion but Nature cries out against slavery and the slave trade.' A moment's thought will indicate that a thriving 'public opinion' is needed to turn round an all-pervading feature of life such as slavery, torture or the maltreatment of prisoners. Again and again the will of the despot, however benevolent, is insufficient to ensure observance.

14 Modern Haiti consists of the western third of the island of Hispaniola, the second largest West Indian island after Cuba. The eastern two-thirds forms the Dominican Republic. When Columbus arrived from Cuba in 1492 the island as a whole, Hispaniola or San Domingo, was densely inhabited by native Amerindians who, the Spanish claimed, were feeble-minded and physically weak. The Indians were replaced by negroes from Africa, and sugar was introduced, rapidly becoming the staple crop. Haiti became French in 1697, and after 1740 developed into a very prosperous and successful sugar-producer, with better land and more opportunity for irrigation than elsewhere in the Caribbean. By 1780 Haiti was the greatest sugar-producer in the world, but because of the activities of the Royal Navy had difficulty in getting its produce to Europe in wartime. In 1793 the population consisted of a privileged caste of white planters, black slaves and free half-castes who, however, lacked civil rights. The negroes were over-hastily emancipated by the French Assembly in 1790 in accordance with the best revolutionary principles: the whites objected; the half-castes havered. Chaos and much cruelty ensued, and the French metropolitan government intervened. So did the British, who after 1794 were at war with the French. The most unspeakable atrocities of every kind were perpetrated by all sides. A great black leader, Toussaint l'Ouverture, appeared and ruled by ancient, animistic means. More economic chaos resulted, and most of the whites left for Cuba, Martinique and Guadeloupe. From being the greatest sugar-producer in the world in 1790, the colony degenerated into a voodoo-ridden 'free' slum, without economic importance of any kind, and with little political freedom. Thus has it remained.

15 The population of Somerset, in 1660 a fully settled English agricultural county with a thousand-year history, was at that date about half the density of Barbados, in other words 120 to the square mile. This included the city of Bath, with a population of 5000 or more, and five or six other towns with populations of 1500 or more, besides over 400 villages with an average population of about 150–300. The total, 200,000, was about one-third of today's population. The population density of Barbados was probably only exceeded, if at all, by that of parts of the Netherlands or the Po valley in northern Italy. The landed population of little Barbados was absolutely greater than that of huge Virginia until after 1680.

16 'Seasoning' (which would today be called acclimatization) involved certain severe changes in the negroes' lifestyle. The diet consisted mainly of cassava and maize – it was better than that on board ship, but not so good as in Africa. New diseases, of which the worst was yellow fever, threatened them. The work was new to them, the 'education' very painful and the overseer almost certainly a brute. Curiously, men survived seasoning better than women.

17 Though the black slave suffered from psychological morbidity compared with the more cheerful nature of the indentured white, the black had significant health advantages. The pure negro's high susceptibility to the inherited disease of sickle-cell anaemia makes him immune to malaria. However this medical advantage did not compensate for the hopelessness of the negro's outlook. No one in the Caribbean appeared to have thought of manumission (emancipation) as a policy. After all, if the expectation of life was only ten years, then manumission at that time might well have improved output meanwhile. This solution was dismissed in Virginia in the 1780s as being impractical because of what was considered to be the essentially childish nature of the black.

18 Dr Johnson.

19 Harold Perkin, *The Age of the Railway*, Penguin, 1970

20 Jeafferson (a West Indian resident), letter of 1668, quoted in Deerr.

21 Much skill and even more luck were needed to make a reasonably rapid Middle Passage, a 4000-mile leg. Recorded voyages took as little as twenty-three days and as much as ninety-five days. In the latter case, there was more than a taint of cannibalism attached to the survivors.

22 To be fair, some captains were identified whose losses ranged from nil to 5 per cent in a lifetime of voyages. Others, less humane, efficient or patient, would regard a 25–40 per cent loss as normal. In the end, the competent captains were employed on their record, and the more stupid, brutal or bungling found themselves out of a job. Almost universally, the correlation is with the individual captain and crew. No other causative factor can explain the very wide differences in the level of loss.

23 Because of imperfect communication, it was assumed until after 1815 that all ships 'below the line' were enemies. The 'line' was the Tropic of Cancer, 22½°N. This situation resulted in several bizarre incidents of single-ship actions between French and French, Dutch and Dutch, and English and English, both ships in each case believing the other to be an enemy.

24 Table derived from Bryan Edwards, *West Indies*, 1819.

25 The mercantile system grew slowly after the Renaissance and the great European expansion of the early sixteenth century; it was probably the mirror of Venetian experience in the Middle Ages. After it developed, it became a doctrine which stated that gold and silver – specie – are wealth as valid as land; that there is more value in a dense population engaged in working up manufactures than in a modestly settled people producing food or raw materials; that exports are more valuable than imports, since saving is worthier and more profitable than consumption; that saving, exports, acquisition of gold and the use of a country's own ships should be deliberately encouraged by state action. Mercantilism was adopted by the Dutch, the English and the French, in that order. The English Navigation Acts, instituted by Cromwell, gave a tremendous boost to the power of the state to intervene as well as impetus to help develop trade and empire. Mercantilism was succeeded in the nineteenth century by free trade, a strong incentive for which was distaste for the 'Big Daddy' attitude of government that was necessary to ensure the success of the mercantile theory. Carried to its logical conclusion, mercantilism was unfit for modern life after about 1720, though it lingered on in Russia, for example, until neatly overtaken by the revolutions of the twentieth century. Even today, many European socialists – the British Labour Party, for example – are still mercantilist, largely because they believe in 'planning'. Adam Smith 'proved' in the 1780s that no man, however wise, can plan as well as the market when freely functional. The problem with free trade has nearly always been unequal distribution of information, ability, wealth and opportunity. This makes 'free trade' less than free.

26 If the British, with their naval supremacy, abolished the slave trade, the rest of the world lost nothing by following suit, since they could in any case no longer pursue the trade.

27 Dynamic state control.

28 'Jerked', as in 'jerked (dried) beef', is one of the few words of Carib origin in the English language.

29 The importance of leather in Europe and the USA before the invention of vulcanized rubber by Charles Goodyear in the mid-nineteenth century cannot be underestimated. Besides its obvious use for boots, shoes, saddlery and harness, leather was also used in the same way that paper, rubber and plastic are today. It was essential for the production of protective clothing, bags, books, boxes, vessels of all sorts, buckets, bellows, washers, seals, pumps and belts in all kinds of machinery. Huge areas of the world were devoted to the raising of cattle or sheep whose meat and bones were left to rot on the pampas, steppes or veldt. The demand for leather was apparently insatiable, and shortage sometimes imposed a restriction on industrial and economic progress.

30 Foreign Office despatches, quoted in Hugh Thomas, *Cuba*.

31   ***Chronology of the abolition of slavery***
   1761   Slavery abolished in mainland Portugal.
   1775   Slavery abolished in Madeira.
   1789   First abolition effort by Wilberforce in House of Commons.
           Slave trade forbidden to Danish subjects as from 1 January 1803.
   1794   Freedom of all slaves decreed by French Assembly.
           Export trade in slaves declared illegal by Congress.
   1800   US citizens forbidden to engage in slave trade from and to foreign
           countries.
   1802   Slavery reintroduced into French colonies.
   1807   Abolition of inter-state slave trade within USA, 2 March.
           Slave trade illegal for British subjects from 1 March 1808.
   1811   Slavery 'abolished' in Spain and Spanish colonies. Violent opposition
           in Cuba. Motion not enforced.
   1813   Slave trade abolished by Sweden for Swedish subjects.
   1814   Slave trade abolished by Netherlands.
           United States and United Kingdom agreed at Treaty of Ghent
           to co-operate towards suppression of slave trade.
   1815   Portugal agreed to abolish slave trade north of the Equator.
           France agreed to abolish slave trade in 1819 (later extended to 1830).
   1816   Slave-owners of Ceylon agreed to free slaves born after August 1816.
   1818   Slave trade abandoned into Dutch East Indies.
   1820   Slave trade abolished by Spain.
   1824–40   Abandonment of slavery by most of new Latin American
             republics.
   1830   End of French slave trade, but beginning of system of *engagés libres*.
           These were bought or kidnapped in Africa, transported to the
           Americas, and became free on arrival: no better, no worse, than
           forced labour. This system lasted thirty years.
   1833   British abolished slavery as from 1 August 1834, with five years'
           'apprenticeship'.
   1842   Emancipation in Uruguay.
   1843   Emancipation in Argentina.
   1843   Slavery abolished in British India.
   1847   Slavery abolished in Swedish colonies.
   1848   Emancipation in French and Danish colonies.
   1852   Last legal importations of Africans to Brazil.
   1854   Abolition of slavery in Peru. Peonage substituted.
   1858   Portugal abolished slavery in colonies to take effect after twenty years'
           'apprenticeship'.
   1860   Abolition in Dutch East Indies.
   1861   Russia abolished serfdom. Emancipation of slaves in Dutch West
           Indies.
   1865   13th Amendment to US Constitution formalized abolition of
           'involuntary servitude' except for criminals.
   1870   Emancipation in Spanish colonies.
   1871   Abolition of slavery in Brazil, to be effected by 1888.

32 A mulatto is a half-caste, with one black and one white parent. A quadroon is
the product of a white and a half-caste, i.e. one-quarter negro. A sambos or zam-
bos (from the Spanish *zambro* = bow-legged) is a three-quarter negro, the child of

a black and a mulatto; hence, perhaps, 'Sambo', originally a black servant, and in turn 'little black Sambo'. There are very few pure negroes in the Americas, and there used to be fewer every year, as more and more 'passed' as whites. At one time as much as 10 per cent of the negro population of the USA was said to be 'passing'. This no longer applies today, because blacks have become proud of their racial origins, and/or taken their parentage for granted, and/or seen themselves as people, not as stereotypes. It is worth emphasizing the difference between blacks and negroes, though its importance is not recognized in the USA. All negroes are black, but not all blacks are negroes. The definition of a negro in this book is a human being with sickle-cell anaemia (see note 26 to the chapter on Quinine). This anaemia does not always occur in half-castes, and very rarely in quadroons.

33   The idea of a central mill was not original. It was first proposed in Cyprus in the fifteenth century in a deal which involved the Venetian family of Martini as bankers. Even then, it may not have been the earliest known example of this particular industrial rationalization. Some mention is made of central mills supplied by barge on the Ganges in India in 200 BC. But the employment of slaves and the difficulty of rationalizing their organization in large numbers made any movement towards sensible economies of scale (central mill apart) almost impossible.

34   American policy in the twentieth century had to take account of four competing interests: US beet-growers; US cane-growers, limited to Hawaii, Florida and Louisiana; privileged Caribbean producers such as Puerto Rico, the Virgin Islands and Cuba; and others. 'Others' found no room in the US market. Cane production in countries such as Cuba, with low wage levels, was much more profitable than at home. Domestic beet-producers could obtain the help of at least forty senators from twenty states, and to the influence of the beet-grower were added the voices of the ancillary industries and the cattle barons, fattened by the beet residues. Beet also involved water rights, a profitable stalking ground for statesman and lawyer alike.

35   The 'Good Neighbour' policy was a strategic/political/economic development idea of President Roosevelt's administration in the 1940s. Before 1939 South America was virtually an Iberian/British sphere of influence, because the Monroe Doctrine of 1823 (aimed at keeping European influence out of the Americas) was dependent upon the supremacy of the British (not the US) Navy to keep the Europeans out. In the 1930s Italian and German influence in Hispanic America prospered notably, while the indigenous governments showed no sign of increased stability. Indeed Argentina, the sixth richest country in the world in 1929, had declined to forty-third place by 1950, largely due to maladministration. In order to prevent a Fascist strategic takeover, and to replace British power, then wholly occupied elsewhere, Roosevelt effectively reinvigorated the Monroe Doctrine, offering some sort of economic advantage to the Hispanics and adding a great deal of rhetoric. Effectively, Latin America was kept at least nominally friendly to the anti-Fascist cause, while the various regimes of varying degrees of corruption made good use of American capital for their own purposes. Little permanent good was done to the 'ordinary people' of most of the countries involved.

36   To the fury of Southerners, every North American is called *Americano* or *Yanqui* in Iberian America.

Tea

Tea
and the Destruction of China

In the 1770s a Mr Twining, head of a firm of tea importers which still carries his name, wrote a pamphlet in which he claimed that there was a village near London whose primary product was material for adulterating tea. The village produced 20 tons of this material a year, and sold it to the trade at half the going price of tea itself. The adulterants were 'ash leaves, collected by children and boiled in a copper with sheep's dung. The mixture is then trod upon to exclude the water, dried, and carefully roasted till the product resembles tea leaves. . . . For scented teas of a finer nature, the children are set to collect elderberry flowers, which are dried and roasted and sold at twice the price. . . . '[1] More generally available, and more generally used, even in the memory of people still living today, were iron filings. The fact that adulterants were such big business shows just how necessary a commodity tea had become to the British in the hundred-odd years since its introduction.

Tea, coffee and cocoa all arrived in London in the same year, 1652. The word 'tea' occurs in Shakespeare, and 'cha', the Canton-Macao form, crops up in Lisbon from about 1550. 'Cha' was used in English seaports until quite recent times, and was corrupted inland to 'char' (no link with 'charlady').[2] The upper classes pronounced it 'tay' and spelt it 'thé', as in French, and this was the common pronunciation of the tea-drinking classes until Victorian times. The word was also spelt 'te', 'thia' and 'kia', in imitation of various Mandarin alternatives in China itself. Significantly, the word was unknown in India until it was introduced by the English.

The Portuguese were probably the earliest tea-drinkers in Europe, since they brought it to Lisbon from about 1580 onwards. They may also have been keen consumers of the Arab mint tea, which was a well-known infusion before the arrival of tea itself. Today we are offered chamomile, linden flowers, comfrey and many other herbal teas, yet there is no evidence that any of them, including mint tea, was ever widely drunk. Mint was only used in Moslem countries, where alcohol was not permissible, or by people who had acquired the habit from Arabs. So why did tea become so popular in Europe?

The reason is that all successful non-alcoholic drinks contain socially acceptable stimulating drugs. Without them, they are no more effective than hot water. In the nineteenth century organic chemists finally identified the alkaloid of tea, theine, as being the same as that of coffee, caffeine. To be fair, theine is less effective.

At the time of the introduction of these beverages water was not really fit to drink in most towns and villages, unless one knew how and when and by whom the water had been drawn from the spring. So, to avoid the risk of waterborne disease, people had to drink safe, boring, boiled water, or alcohol strong enough to kill germs. A simple method of allocating importance to a resource that modern man takes for granted is to imagine life without it. Non-alcoholic beverages are an excellent example; it is hard to see how one could manage without them. The alternative, alcohol, however mild, must have had an unfavourable effect upon the conduct of business, the schooling of the young, the control of horses or the navigation of a ship. It is difficult to imagine a modern world without stimulating but non-alcoholic drinks, and the demand for tea after its arrival in Europe was evidence that it filled a need.

By 1820, millions of pounds of tea were being imported into Europe every year, and re-exported all over the world, more than half by the British. Probably 30 million lb was consumed in the United Kingdom alone. Despite its high cost at this date tea was drunk throughout the British Isles by all who could afford to buy it.

All the tea came from Canton, on the southern coast of China, and none of the merchants had as yet penetrated inland. Although China had made available to the West much of the technology of the pre-industrial age, tea-growing and tea-curing had never moved westwards, only eastwards and southwards to Japan, Formosa and Java. Tea was unknown in India, except as an imported consumable from China, enjoyed only by some Europeans and a few Europeanized Indians.

History's joke on Europe is that for nearly two centuries a commodity was imported halfway across the world, and that a huge industry grew up involving as much as 5 per cent of England's entire gross domestic product, and yet no one knew anything about how tea was grown, or prepared, or blended.

In that great burst of mercantile activity which followed the Renaissance, western European influence spread about the globe and created the world we know today. But Europe held many beliefs which we now know to have been fallacies. It thought of itself, for instance, as superior, gifted, in a position of natural leadership. However, there was no technology known to the Renaissance Italians of which the Chinese of the same period were not also aware (see table on p. 96). A great deal of highly developed intermediate technology in fact put China and Japan, and in some instances India too, far ahead of Europe. Persia and some of the Arab-controlled countries were equally in advance of much of Europe. The one exception to this was gunpowder, discovered by the Chinese, who abhorred its use for violence or warfare, as did the Japanese.

It is entirely possible that the Chinese rejected the use of gunpowder for violent purposes not because they were noble lovers of fireworks, as opposed to gunnery, but because they realized that guns would destroy the complicated feudal system which they had evolved. This view is certainly sustainable in Japan.[3] But whatever the reason, because the Orient denied itself the use of gunpowder for violence it laid itself open to defeat by the Western barbarians. With guns, European ships could survive at sea, European armed men could establish themselves ashore, and European armies could ultimately defeat the natives, even if those armies contained many native soldiers. Guns of all sorts, gunpowder and the precious metals were the only objects which could be generally traded in the Far East. It was not a backward, empty part of the world, like America or Africa. East of Suez the ruling classes were sophisticated in the best sense, highly educated, of a far older civilization than any in Europe, and clever in their dealings with other peoples. To any shipmaster or trader who achieved the long voyage, the merchants of the East must have appeared as magical and complex as any great man in any great city in the West.

The second mistaken view was that most people in the world, except for the post-Renaissance Europeans, were incapable of rational thought. This form of invented fallacy is akin to that which says that women cannot think straight, or that children cannot pursue an idea beyond the same evening, or that all animals are incapable of elementary reason of the cause-and-effect kind. Indeed, it was common even in the early twentieth century to believe that Orientals were so mixed-up and mysterious that their thought processes were quite different from those of Western man. Certainly, the sixteenth-century Chinese were capable of the subtlest kind of reasoning of a Jesuitical nature, and they appreciated and made use of Jesuits for special diplomatic purposes.[4] The Chinese and Japanese also produced the most elegant engineering designs, using such materials as were available in each locality. Before 1700 Japan, China, India, Persia and the Arab countries were certainly more advanced in their knowledge of the natural sciences than most Europeans, and for this to be true the reasoning powers of the elite must have been of a high order.

Technological advances transmitted from China to the West

Invention or discovery	First known date (AD unless stated otherwise)	
	China	**Europe**
Rotary winnowing machine, with crank handle	40 BC	Late 18th century
Rotary fan for ventilation	180	1556
Blowing engines for furnaces and forges, with waterpower	31	13th century
Blowing engines for furnaces and forges, crank-drive type	1310	1757
Draw loom for figured weaves	*c*.100 BC	4th–5th century
Silk-working machinery:		
reeling machine	1st century BC	} both *c*. end of 13th century
flyer; twisting and doubling	1090	
waterpower applied	1310	14th century
Wheelbarrow	231	*c*.1200
Iron casting	2nd century BC	13th century
Concave curved iron mouldboard of plough	9th century	*c*.1700
Seed drill plough, with hopper	85 BC	*c*.1700
Gimbals	180	*c*.1200
Shipbuilding:		
stern-post rudder	8th century	1180
watertight compartments	5th century	1790
Rig:		
efficient sails (mat-and-batten principle)	1st century BC	19th century
fore-and-aft rig	3rd century	9th century
Gunpowder:	*c*.850	13th century
rockets and fire lances	*c*.1100	15th century
projectile artillery	*c*.1200	*c*.1320
explosive grenades and bombs	*c*.1000	16th century
Magnetism:		
floating magnet	1020	1190
knowledge of magnetic declination	1030	*c*.1450
theory of declination discussed	1174	*c*.1600
Paper:	105	1150
printing with wood or metal blocks	740	*c*.1400
printing with movable type	1045 (earthenware)	–
	1314 (wood)	–
printing with movable metal type	1392 (Korea)	*c*.1440
Porcelain	3rd–7th century	18th century

Adapted from Needham, *Science and Civilisation in China*, Vol. I, table 8 (CUP).

Finally, it is a great mistake to believe that information available to the Portuguese was also known to the Dutch, or that the French and English shared a common pool of received wisdom. Sailing directions, charts, positions of snags and rocks and reefs and eddies and currents – information which the world now shares – were then jealously guarded secrets. If Magellan managed to sail through the Straits to which he gave his name in 1519, or if the French managed to go round the Horn earlier and with more ships than any other maritime power in the seventeenth century, or if the British had a tradition of dead reckoning by longtitude in the eighteenth century which made their navigators the envy of the world, it is wrong to believe that the knowledge or advantages which enabled them to make these achievements became quickly available to other nations. These were practically state secrets, and to reveal them to foreigners was to court justifiable execution.

It follows that later explorers or voyagers from Europe did not start out with obsessions. They did not, like Columbus, die imagining that they had discovered Asia when they had found the Caribbean. They did not resemble those who sought the Northwest Passage, like Hudson. They had little in common with those Spaniards who believed in Eldorado. The prudent behaved in a much more pragmatic way, seeking knowledge by navigation, enhancement by exchange and profit by piracy. The realities of life at sea and the unknown nature of the lands they visited created a new race of mankind that followed the rules of expediency. The same spirit which permitted astronomers of the age to reject the conventional view about the sun rotating round the earth also made seamen contemptuous of land-based authority. The post-Renaissance exploratory phase of the European adventure abroad produced daily evidence to acute-witted travellers to the east, south or west that conventional wisdom was of no value to men in their situation. Was this a concept of the same kind as that which occurred to the religious reformers? Or did the religious reformation precede the great European oceanic adventure? Whatever the answers to these questions, successful Mediterranean powers have always been able to comprehend conditions all through that sea. The Atlantic presents the antithesis: no Atlantic power has ever succeeded unless it has left the man on the spot to his own devices.

This simple fact is reinforced by the tyranny of the time/distance equation, which did not materially improve at sea before about 1830–50 and which has certain obvious effects.[5] The man on the spot has to have the appropriate qualifications, or the operation fails; and the powers at home need the ability to select those who can succeed without reinforcements of intelligence, money or manpower. No political system ever produced the right answer, but the ultimate primacy of the British in East–West trade can be traced to the fact that in following these 'Atlantic rules' the British were usually more successful than the opposition.

The true British failures, such as the loss of the American colonies, can be blamed on the Atlantic rules being broken by an ambitious set of politicians trying to serve a king who did not understand the peremptory nature of this logic – a logic which also made the colonist far freer than anyone at home. Even the feudal Spaniards were quite clear as to why men went to America. The conquistadores said that they had volunteered to serve their king and their God and – a pause – to become rich. They were obsessed by gold and silver, and regarded the huge Spanish Empire as a fief of the king. The Portuguese regarded themselves primarily as trading merchants rather than imperialists, as did the Dutch and English in their turn. All three peoples tumbled into foreign possessions as an adjunct to trade, if not, as is often said of the English, in a fit of absent-mindedness. The French position was somewhere between that of Spain and that of England, with the State playing an important part, as in Spain, and more efficiently, but the individual also enjoying more opportunity than in Spain.

The post-Renaissance exploratory phase was rather like the Crusades in that western Europe was on the move; but instead of an immense effort which achieved very little, the oceanic efforts of western Europe changed the world. The difference arose partly because the horizons were so much wider, partly because the times had changed, and partly because Atlantic peoples and not Mediterranean ones were involved at the core; but largely because the individual, even in rigid Spain, was able to win enough wealth to buy himself an estate at home, a title perhaps, and a noble bride, and to found an aristocratic family.

This opportunity for socially upward mobility was of fundamental importance. For the first time in history a crude, coarse seaman could earn enough by means close to piracy to establish himself among the elite of a stable society. Piracy could be practised abroad and when they returned home the survivors would not be averse to keeping silent, or to painting a rosy picture of what had occurred, or to suffering selective amnesia. The few who had weathered Hawkins' attacks on the Spanish Main, or Drake's circumnavigation of the globe, would either lose their fortunes in a few weeks or months, or settle down at home several notches higher on the social scale than when they had left. In such ways were the elite of society selected.

The immediate cause of western European Atlantic expansion was the closure of the Eastern Mediterranean by the Ottoman Turks, served by Greeks, who defeated every rival until the battle of Lepanto in 1572. Trade was inhibited after the fall of Constantinople in 1453, and became particularly difficult from the 1480s onwards.

All the early westward transatlantic voyages were aimed at the discovery of a route to the Indies – originally a blanket word for India, Malaya and all the East Indies – from where most of the spices came. Following Vasco da Gama in 1498, in the next century some two hundred eastward voyages were undertaken round the Cape of Good Hope; most of them were also Portuguese, and perhaps fifty were originally Dutch.[6] Only half the ships returned.

When the Portuguese and then the Dutch finally reached the source of the spices, the products first became much cheaper in Europe, and then each country in turn tried to establish a monopoly. The means of trade were East India Companies, bodies of merchants who came together to reduce risks and internecine competition, and in the hope of becoming stronger and more successful than they could on their own. They used bigger ships with better crews and more armament, and their record of achievement is generally better than that of individual merchants.

East India Companies were established by the English, Dutch, French, Danes, Spanish, Swedes, Scots, and for a very short time the Austrians. The earliest was the English: founded in 1600 and known colloquially as 'John Company',[7] it was ultimately to be the most important.

The Dutch had been the traditional northern European carriers of the world's tropical products from Lisbon, the port to which the Portuguese brought spices from the East until 1580, when Spain and Portugal were 'united'. Lisbon then came under the deadening control of Madrid. The Dutch found prices rising against them; corruption and inefficiency ruled in place of enterprise; trade was subordinated to the missionary fanaticism of Spain; war became almost continuous in the Netherlands and on the high seas. To trade directly with the East Indies was the only answer.

By the end of the century the merchants of Amsterdam, Zealand, Delft, Rotterdam, Hoorn and Enkhuizen were sending more than a dozen ships a year to India, the Malay archipelago and the East Indian islands. They fought the Portuguese, the English and each other. Logic dictated a national combine, and so the Dutch East India Company was formed, with a European headquarters in Amsterdam. Trade was regulated; customs duties were removed from the Company's imports; it was authorized to make treaties, maintain armed forces at sea and on land, set up trading posts (called 'factories') and coin money.

The local headquarters was at Batavia in Java, on the site of the ruined native capital of Jacarta, today restored in importance and known as Djakarta. The Dutch expelled the Portuguese from Ceylon and Malacca and established the first white colony in South Africa, with momentous consequences. Ultimately they had eight foreign 'governments', in

Amboyna, Banda, Ternate, Macassar, Malacca, Ceylon, the Cape of Good Hope and Java. Factories were established in Bengal, Coromandel, Surat, Thailand and the Persian Gulf. Until about 1670 the Dutch East India Company was the richest corporation in the world, supporting the high Dutch civilization which produced Rembrandt, Vermeer, Frans Hals, Vondel, Grotius, Spinoza, the greatest printing trade in the seventeenth-century world, together with innumerable unsung minor writers, poets, painters, architects and, above all, patrons. Before the decline of its powers the Dutch East India Company mustered 150 trading ships and 40 ships of war, with 20,000 sailors, 10,000 soldiers and nearly 50,000 civilians on its payroll, and still managed to pay a dividend of 40 per cent. The Company was the envy of all its rivals.

The Dutch traded between Japan, China, India, the Persian Gulf, Africa and Europe, and between all those places and Amsterdam. Spices were exchanged for salt in the Persian Gulf; salt for cloves in Zanzibar; cloves for gold in India; gold for tea and silk in China; silk for copper in Japan; copper for spices in the East Indies. A huge Indies-to-Indies trade was nearly as profitable as the Grand Trade from the East to Europe. Despite heavy losses from pirates – then, as now, the scourge of the China Seas – weather, European rivals, corruption, inefficiency, theft and disease, the Company prospered. The Dutch were without scruples: they raised the price of essential spices by 180 per cent; they formed monopolies; and they destroyed local rivals.

The Dutch and English had been ideological and religious allies against Spain at the time of the Armada in 1588, but became trade rivals a few years afterwards. During the Thirty Years' War of 1618–48, which devastated large parts of central Europe, both England and the Netherlands tried to keep aloof from this vicious, internecine and seemingly interminable war between Christians, which was fought nominally for faith, not for political or national advantage. Anglo-Dutch rivalry was expressed on the high seas, not in Europe.

The first Anglo-Dutch combine, that of an East India Company, had been proposed in 1613, but the men on the spot refused to ratify the merger which had been almost agreed between Amsterdam and London. A solemn 'treaty of defence' was signed in 1619, but when the news reached the East a kind of farcical joust ensued. For an hour, hostilities ceased: Dutch and English ships dressed their ships overall with flags and sailed up and down, saluting with blank shot. After an hour the flags were taken down, the ships stripped for battle, and the cannon loaded with shot once again. The undeclared war culminated in 1623 with the Amboyna massacre, in which nearly a hundred men, women and children of the English East India Company were tortured and killed. These final atrocities effectively

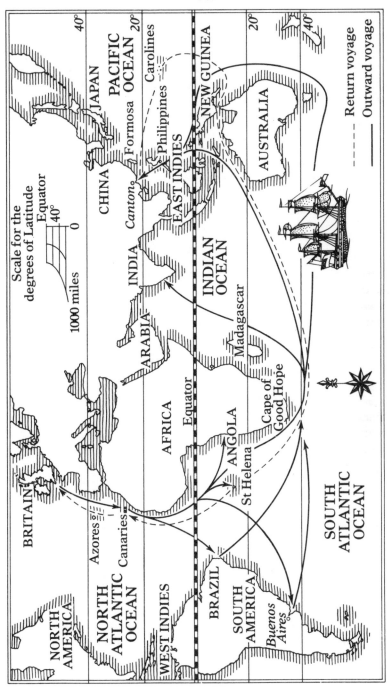

The Routes Taken by the East Indiamen

confirmed Dutch hegemony in the East Indies; the English were confined to the mainland of India itself.

The far-reaching consequences were not immediately apparent\to the English, who nurtured a slow, angry resentment which was to mature in the wars against the Dutch in the times of Cromwell and Charles II a generation later. General Monk, a moderate Protestant and no lover of things Spanish, quoted with approval the Andalusian proverb: 'Revenge is a dish which tastes more pleasant when cold.' Only the combined monarchy of William III of Orange and Mary II of England from 1689 was to bring Anglo-Dutch rivalry in the East to an end, and by that time the Dutch were in decline. Their trade was now reducing every year, confined to the Grand Trade between the East Indian islands and Europe, and their East India Company was bogged down by bureaucracy, corruption and overheads, and enervated by monopoly profits. Driven from their trading preeminence in the Indian Ocean, the Dutch were to survive as colonists in the East Indies until in World War II the Japanese demonstrated to the indigenous Indonesians that the White Ruler was not for ever.

The English East India Company owed its success to three factors of an accidental nature. Like other happy accidents in history, these discoveries became 'principles' by which men subsequently conducted their affairs. The earliest, and perhaps the most important, of these accidents took place in 1609. To help alleviate the slump and lack of purpose in the shipbuilding trade under James I, particularly in the southeast of England, the Company decided to establish its own shipyard on the lower Thames. Thus were developed the East Indiamen, the best merchant ships in the world until the arrival of the American clippers in the nineteenth century.

The English East Indiamen had to contend with Indian, Chinese and Annamese pirates; European privateers, often indistinguishable from pirates; Portuguese and French, as well as Dutch, rivals who, it might be discovered at the end of a voyage lasting more than six months, were now at war with England. Sometimes a foreign ship would pretend to be a pirate, and turn out to be a naval vessel of another country. Sometimes a ship would pretend to be a naval vessel and turn out to be a pirate – Chinese and Japanese junks were very fond of this ploy. Sometimes a brigand–ruler of an island would be encouraged by the Dutch, Portuguese or French to board an East Indiaman in a friendly manner, and then try to seize the ship – junks in the East Indies were inclined to this activity.

For all these reasons the East India Company built their ships well, fitted them with the best sails and rigging, and finished them with tar and paint of the highest quality. Because they had to carry bulky cargo the ships were, to modern eyes, massively plump and awkward to windward; but they were of finer design and construction, it was said, than those of any navy, including that of France which was pre-eminent for long periods. The East Indiamen could not outsail pirates or other enemies, so they were supplied with enough powder and shot to fight at least two battles, and although they often sailed in convoy, there are many records of successful single-ship actions.

They could renew their stores at St Helena, a British base for fresh food and water and other necessities of life at sea, and an absolute possession of the East India Company from 1674 until the Napoleonic Wars, but the Cape of Good Hope was in the hands of the Dutch, who were not always friendly, while Madagascar was a fierce, inhospitable, little-known island, a nest of pirates of all European nationalities. Ships frequently called at the Canaries to take on stores of citrus fruits as an ascorbic against the dreaded scurvy, but would then often travel down the length of the Atlantic, round the Cape of Good Hope, and all the way to India or China without seeing another vessel.[8] This 25,000-mile voyage took perhaps six months. Other ships might have an even longer trip if chance, wind or weather obliged them to call not only at St Helena and the Canaries, but also in Brazil, Buenos Aires or Angola. The intention was for the ships to be so well found that they were self-sufficient.

As with ships, so with men. 'Freight is the mother of wages' was a doctrine which held good in English law until 1854. A shipwrecked mariner was not rewarded for his unsuccessful voyage, and a man was only paid at the end of a complete round trip, England to England. Men had to be selected and chosen for their ability to work and fight a ship with a far smaller crew than a vessel of equivalent size and armament in the Royal Navy. In modern parlance, productivity had to be high, and the men therefore of a corresponding high quality. East India Company ships attracted better officers than did the Royal Navy, and the men were not forcibly pressed, as in the Navy, but volunteers. Adventure, a desire to see the world and a love of the sea were, as always, the reasons for young men to go to sea, but a particular motive for joining the East India Company was the attendant exemption from impress, kidnapping, conscription, hijacking or whatever it might be called by the Royal Navy, at sea or on shore. As a result the crews were of the best, not the paltry collection of men of all nations, the dregs of the world's seafarers, with which the Navy had to do its business.

The Company had also, by chance, managed to make the interests of its officers coincide with those of the management, by giving the ships' officers a right to a certain free tonnage with which to trade. This privilege varied from time to time, but might be about 10 tons for a captain, with other officers pro rata. The profit on a ton of tea was equivalent to a year's wages if it were magnified by evading customs duty and selling the tea to smugglers, a common enough practice. Because the privilege of 'private trade' was limited to space, not weight, the tendency was for officers to buy or to transport only the very best tea. Smuggled tea therefore acquired a reputation of being of higher quality than the legally imported supplies auctioned by the East India Company. The Company, of course, frowned on smuggling, and the crew, if maltreated, would inform on any captain and officers who engaged in the practice. But there was enough profit in the smuggling transactions to grease the palms of most of the crew. This gigantic conspiracy was partially stopped by the reduction of tea duty in 1784, but other goods took the place of tea and 'private trade' was always a very important element in the saga of the Company.

For good or ill, China turned her back on the world during the fifteenth century. The Court was transferred from Nanking to Peking, much further north. Overseas trade by Chinese nationals was gradually abandoned, until in 1521 it was declared illegal. The myth was established that China lacked nothing. From about 1500, the Chinese genius for improvization, development and philosophical enquiry was replaced by a static defence of the past, a regressive complacency about the superiority of Chinese culture, and a refusal to accept the worth of anything foreign. This psychological defence was matched by a requirement that foreign trade, if any, should be conducted by foreigners, though the Chinese were sometimes still the actual negotiators, bankers and merchants, known as hongs. Foreigners were despised as 'foreign devils', but for reasons of state trade was carried on by Japanese, Koreans, Formosans and Europeans.

Before 1840 there were no treaty ports in China; Europeans were restricted to a small enclave at Canton, and permitted to trade only with the hongs, who were sureties for payment of customs dues and for the proper conduct of business. The trade in China tea was largely in the hands of the British East India Company, who had established a near-monopoly in the finished product, challenged by the Dutch, Portuguese and French with varying success. This was the position from about 1686 until 1834, when the liberal sentiments of the day led the English to throw open the tea trade to all comers.

The China ships, which were not only owned but chartered by the Company, sailed direct to and from Canton. During some periods of history they were in fact prohibited from sailing via India, in order to prevent their crews from indulging in a highly profitable local trade or other corrupt practices. On the way out to China round the Cape of Good Hope, East Indiamen often took a track to the south of Australia – in the days before it had been discovered – along the Roaring Forties, and then north past New Guinea. At the end of the eighteenth century they returned from China almost due east from Canton, through the passage between Formosa and the Philippines, then south through the Carolines, to Dampier Strait at the northern end of New Guinea. This track, which left the Pacific north of Australia and was fully 4000 miles longer than going via Singapore, took less time because the winds were usually fair. It was also chosen to avoid unfriendly ships in the narrow waters of the Straits of Malacca.

In the century and a half after 1680, almost no one in the Western world knew anything about tea: what it was or how it was grown, treated, sorted and blended. Their acquaintance with it began on the wharfside at Canton, and, as long as the quality of the product could be ascertained by a ritual tasting, and the price was reasonable, the European buyer was satisfied. An elaborate chain of intermediaries was imposed between the Chinese peasant who grew the tea and the European who drank it. There was a tea purchaser in each village who might buy the few pounds each peasant produced at each plucking. A tea centre in each district prepared the tea for sale, after which it went to an elaborate provincial sorting and blending establishment. The tea was then sent to Canton by water, by packhorse or on the backs of coolies; here the cases were opened and the whole process of blending began again.

Much adulteration took place along the line. After arriving in London, Amsterdam or Paris, the tea was sorted, blended and adulterated again. Then it was auctioned, and the merchant would blend and pack the tea in quantities small enough to sell to the shopkeeper, who might blend and adulterate again. Tea is a dried and blended preparation of the leaves, buds and flowers of a single species of camellia, *Thea sinensis*,[9] and both Europeans and Chinese adulterated it with twigs, both of the tea camellia and other members of the Theaceae family. They also used wood, pine bark, alien leaves, sawdust, soot and Prussian blue, as well as legitimate additives in the form of scented bergamot, orange, lemon, verbena or other shrubs to give the tea a particular taste.

This question of adulteration was one of the early answers to the exponential European demand for tea which the Chinese were unable to satisfy legitimately, and it illustrates the innocence of the European buyers. The merchants did not mind adulteration as long as they could pass the product on and there was no customer to gainsay them, and when the Europeans began to apply rational methods to the problem of producing tea they discovered that the Chinese had been less than efficient for nearly two thousand years. It is one of the great ironies of the trade that the bergamot-flavoured Earl Grey, now regarded as a 'great tea', was originally scented with *Chloranthus inconspicuus, Jasminum sambac, Gardenia florida, Murrya exotica* and *Aglaia odorata*, as well as orange and lemon leaves. The 'great tea' started as a standard adulteration to spin out a shortage.

If properly grown in the right place and properly fertilized, the tea bush will yield up to 5 lb (wet weight) per bush per year. The Chinese peasant grew his bushes as a garden crop, probably innocent of all fertilizer (even human manure), and he might pluck a tenth of what is produced in an Indian plantation today. In the early eighteenth century he would not have been paid more than 1d per pound (wet) for his efforts, equivalent to between 3d and 6d per pound of dry weight. This tea then arrived in London, Amsterdam or Paris, and cost, retail, from £3 per pound in 1700 down to an average of 3s one hundred years later. To this had to be added the exorbitant duty which every European government imposed upon tea imports and, in the case of all continental countries, internal customs duties or octroi as well. From the Chinese peasant to the European consumer there was a mark-up of between eight and twelve times in the hands of the commercial intermediaries, plus the government's impost, which ranged between 50 and 100 per cent of wholesale value and was always levied at the ports. It was no wonder that smuggling and adulteration were common practices, probably doubling the legal tonnage imported.

Legal imports into England were probably about 50 short tons[10] in 1700, at a wholesale value of about £4000 per ton, or £2 per pound. The continent probably imported about a third as much. About a fifth was re-exported by the East India Company to the British colonies, including America. The total imports into Britain grew from 50 tons in 1700 to at least 15,000 tons in 1800. The average for the century was rather less than 4000 tons, and the wholesale, landed but untaxed value in bond was about £350 per ton, falling from £4000 in 1700 to about £200 (or 2s per pound) in 1800. The volume rose 300 times, or 30,000 per cent, while the price fell to 5 per cent of what it had formerly been in London.

All this tea, if tea it was, came from China. All of it passed through the tiny area of Canton in which the Europeans were welcome. Ships would carry only a few hundred tons each, in order to spread risks. Transporting 15,000 tons was therefore quite a task, but a rewarding one.

The British East India Company was believed to add at least a third to the price of tea, thus taking £100 a ton out of the 375,000 tons imported during the century. This global figure obscures the rise, on the same basis, of the East India Company's cut, from a sum equivalent to $17 million at the beginning of the century to an annual equivalent of $800 million in 1800. The East India Company was big business, hated and loathed by smugglers and consumers alike, and a symbol of corrupt, complacent monopoly.

It may be a surprising connection, but the influence of the tea trade upon European porcelain was profound and complicated, and had a great deal more to do with freight and trading requirements and good ship trimming than with mere cups and teapots directly connected with tea consumption. Porcelain is a very fine, translucent ceramic, loosely and originally called 'china' and distinguishable by anyone from earthenware, pottery or stoneware, the crude native products of all other areas of the world before the eighteenth century. The word 'porcelain' itself comes from the Italian *porcellano*, literally a little pig, which describes the shape of the common Mediterranean cowrie shell, whose fine, smooth, translucent surface porcelain was considered to resemble.

Long before Europeans discovered China (or china), porcelain had been exported to Persia, Arabia and Turkey, and fine examples of pre-1500 Chinese porcelain can still be found in everyday use in parts of the East. A hundred years ago an English traveller collected nearly 5 tons of Ming pieces in Persia. When part of the collection was sold on his death in 1905, the sum realized was over £30,000 (nearly $1.5 million in today's money). He had paid less than £200 (£6000 today) for the whole collection thirty years before. Chinese porcelain manufacture was therefore an industry long before Europeans arrived in the Far East. From about AD 800 onwards, Chinese junks or Arab dhows took porcelain all over the Indian Ocean, and shards have been found as far west as Morocco, as far south as Zanzibar and Bali, and as far east as Hawaii. The porcelain had been transported by Chinese, Arabs, Polynesians or East Indians, not by Europeans. The first trade with Europeans was conducted in the Middle Ages in Cairo and other North African entrepots. The first imitation of this china was attempted in early Renaissance Italy; the result was not porcelain, but faience or majolica, because it contained only silica, alumina, lime, magnesium, oxide of iron and carbon in varying proportions. None of the ceramics made in Europe before the eighteenth century contained any kaolin (china clay), the ingredient which is the secret of true porcelain.[11]

The China trade, as conducted by the Portuguese, Dutch and English, in order of sequential pre-eminence, and by the French, Danes and Swedes as minor trading powers, involved two very lightweight, high-value commodities: tea and silk. To trim the ship and make her sail properly about half the weight, but much less than half the volume, of heavy, water-resistant goods or ballast was needed in the bilges. This ballast could be carried permanently in the ship's bottom or externally, as a keel; but permanent ballast paid no revenue – it was dead weight. A far better method was to find some heavy commodity, which could be traded. The problem of ballast was much more acute in the China trade than in any other, since both tea and silk had to be carried in the middle of the ship to prevent any risk of wetting from the sea, from condensation or from rain.

China did have one major raw material deficiency, copper ore, so copper, gold and silver bullion became the medium of exchange with the Celestial Kingdom. Japan supplied most of the copper before the Europeans arrived, after which the various East India Companies came to dominate the copper trade with China. The ships returned with tea, ballasted with mercury, other minerals and porcelain, which Europeans were unable to make until the eighteenth century, as mentioned above. Very roughly, a quarter of all tea imported had to be matched by heavy ballast goods, called 'kintlage' in the eighteenth century; and from the ships' records available, about a quarter of all the kintlage was porcelain. Therefore, for every 100 tons of tea, 6 per cent by weight of porcelain was imported into Europe. Thus if, on average throughout the eighteenth century, 4000 tons of tea were imported in England each year, probably 240 tons of porcelain were also imported. More than as much again would have been imported into Europe and the American colonies.

This huge tonnage of porcelain was often handled by the supercargoes, independent traders carried by each East India Company. The porcelain was treated with scant respect. Heavy objects were wanted, or small, easily stowed plates, cups and saucers. The job of the supercargo was to see that the tonnage was filled. Highly complicated arrangements were very common, and corrupt and convoluted the deals turned out to be: there were orders from England, patterns from France and specifications from the Netherlands. The Chinese hongs in Canton were driven down to the lowest possible prices: £5 10s in 1712 for a 216-piece dinner service; £7 7s in 1730 for a tea service for 200 people, each piece decorated with the arms of the ambassador who ordered it; teapots, 5000 of them in 1732, imported at 1½d each. Even if these prices are multiplied by 100 to give an approximation of today's prices, china of this quality was incredibly cheap.

The tonnage should be put into perspective. A porcelain teacup of the eighteenth century weighed less than 2 oz, and a dinner plate less than 8 oz. There are nearly 36,000 oz to an imperial ton. If the average piece weighed 4 oz, then over 5 million pieces were imported every year into Europe. This kind of order of magnitude is confirmed by the records of certain East India Company ships: 250,000+ pieces in a 20-ton lot in 1718 (average 2.8 oz per piece); 332,000 pieces in a 40-ton lot in 1724 (average 4.3 oz per piece); and 178,000 pieces in 1732 in an 18-ton lot (average 3.6 oz per piece).

The origin of teapots and teacups presents a problem. The Chinese, like many other peoples at a later stage – Indians, Arabs and Turks – did not have pots. Their tea was made in a kettle. The early English silver or other metal teapot was known as a kettle, and the tea was mixed with sugar and, perhaps, other herbs and 'brewed up' in the kettle. This practice is followed in Morocco, Algeria and Tunisia to this day.

Some entrepreneur, observing the beautiful shape of the Chinese wine flask, suggested that it should be copied and sold in Europe as a teapot. Certainly, before about 1720 no factory in Europe would have been capable of making a ceramic pot that could withstand boiling or even very hot water. The pot, therefore, is in effect a foreign invention, not used in China.

The teacup with a handle is also a foreign invention, specifically European. The Chinese did not put handles on their own teacups, because they drank their tea cool enough to render handles unnecessary – though they did not use milk. Handleless cups stow more easily than cups with handles, but handles can be designed specifically for easy stowage rather than easy drinking; some early teacups were designed in this way, and are difficult to use with either delicacy or efficiency. Such cups were designed in England and made in China. Handleless cups were made in England after porcelain manufacture began to imitate Chinese styles. Handles were specifically added to cups made in China after about 1750, and the trade of handle-maker became an acknowledged one in large European cities. About half the cups in use in 1770 in Europe would have had handles, half not. So why did Europeans add handles to cups?

The eighteenth-century Europeans, like the Japanese, but unlike the Chinese or the Russians, regarded tea-making as a ceremony. There was the boiling water, not boiled for too long. There was the specially warmed pot. There was the infusion time. There was the pouring, a little bit of a ceremony all on its own. There was, of course, originally no question of adding milk, either before or after the tea, but there was the problem of eighteenth-century sugar, which had to be put into the cup before the tea or it would not dissolve. The cup of tea was then sipped, hot, and therefore needed a handle. In Russia and China the tea kettle was on the hob all the time, and drunk lukewarm in a small bowl by the Chinese or in a glass by upper-class Russians. Of course, there is no merit in drinking China tea of the highest quality over-hot, since the flavour is best sampled at a temperature a few degrees above blood heat; it should never be drunk at a temperature which requires a handle on a cup. In some societies, praise was showered on the tea by blowing upon the surface and by making exaggerated sucking noises when consuming it. This seems curious, since it would appear to be a criticism of the temperature of the tea, but it should surprise no one: there are a great many variants in the human family's approach to tea consumption.

By the end of the eighteenth century the made-for-Europe trade had come to an end, and the traders had to find something else to act as ballast. At the behest of their own manufacturers European governments had made the export of porcelain for the West a difficult proposition, even at the ridiculously low prices paid in Canton.

Chinese porcelain was a once-fired, hard-paste, high-temperature (1400°C) ceramic, cheaper and better than early European porcelain, which was twice-fired, soft-paste, low-temperature (1250°C) and often misshapen or rejected as imperfect. Chinese porcelain was also stronger, and truly vitrified; the glazes on the European version were often porous or partly porous, more granular and less stable. But from 1760 onwards the Europeans, and particularly the English, developed home-produced stoneware, heavy, coarse, with a short life, but far cheaper even than Chinese porcelain. Hongs were left with enormous stocks of real porcelain on their hands: one merchant in Canton had 12 million pieces left in his warehouse when the British East India Company decided in 1791 to discontinue official imports. The East India Company in London was complaining in the 1780s and 1790s of having huge quantities of unsold porcelain in its warehouse. Private imports by ships' officers as ballast among their tea continued; there must have been many bargains while millions of pieces were being sold off below cost. Twelve million pieces were then worth £20,000 sterling. Today, the same pieces would be worth at least a billion dollars: a good investment which no one made.

If 400,000 tons of tea were imported by the British East India Company in the 107 years between the beginning of its monopoly in 1684 and 1791, when the decision was made to cease imports of porcelain, then about 24,000 tons of porcelain were imported as ballast in the same period. This is equivalent, at an average 4 oz per piece, to 215 million pieces of Chinese porcelain. This sounds a great many pieces, but it only amounts to about five pieces for every Briton who lived beyond the age of ten in the same period. The vast majority of people who used china at all in the eighteenth century would have used Chinese porcelain, because before about 1780 it was cheaper than ordinary pottery. For a short period in history, therefore, the tea trade was responsible for the highest 'quality of life' in the middle-class houses of England.

The tea trade is, of course, responsible for much more than this. The willow pattern was derived in England from a Chinese version of an English idea of a Chinese scene. Chinese images were anglicized: European scenes were chinified. *Chinoiserie*, the concept of European furniture, decoration and textiles imitating Chinese designs, came into fashion as a result of the tea trade. The unlocking of the secret of porcelain manufacture, which took place in Europe between 1709 at Meissen and 1742 at Chelsea, was the direct result of a desire to compete with the Chinese imports, and the product was something different – not strictly comparable in body, glaze, firing or luminescence, but a whole new industry.

Europe's first image of Japan, Korea and Annam also came from the imports of porcelain from those countries; carried by Chinese junk, these goods were traded in the East Indies or exchanged at trading posts in Formosa, whch was occupied by the Dutch for most of the seventeenth century, becoming Chinese only in 1682. Porcelain from these three countries was not as fine as the Chinese product, but better than most contemporary European ceramics: these so-called 'heathen', 'primitive' peoples made a peculiar product in difficult conditions for many hundreds of years before the 'clever' Europeans discovered how to compete. When competition came, it was not exactly on equal terms. The Europeans altered the rules, cheapened the process and changed the product. In 1800 the European pottery industry was only about 5 per cent involved with porcelain; the rest consisted of coarse, heavy wares. This represented a regression in the fifty years since 1750. If the European bourgeoisie had continued to use Chinese porcelain instead of the crude, factory-made earthenware, might not the taste, vision and opportunities of the industrial nineteenth century have been l_ss debased? Or was the loss of Chinese porcelain never regretted by the chauvinist Victorians?

The tea trade was in crisis in the last third of the eighteenth century. All tea came from China and was imported, legally by the East India Companies of the European countries, illegally by smugglers. Over two hundred heavy ships were involved in voyages from Canton: two-thirds of them reached Europe each year. The tea tonnage they brought was probably 12,000 in 1770. English duty amounted to more than half the cost of the tea. It is probable that the illegal trade amounted to an additional 6000 tons. The obvious answer was to reduce the duty.

This proved too simple and too radical a solution for the government of Lord North. For reasons which appeared very attractive at the time, North's cabinet decided to kill three birds with one stone. They would sell the tea in the American colonies, which would get rid of the stuff. They would apply a duty of only 3d per pound (against the norm of 2s 6d, on average, or one-tenth the former duty); this would make the bargain irresistible, and drive the smugglers out of business. Finally, the imposition of this tiny duty, though making the tea irresistible, would compel the colonists to admit the right of the home government to tax the people of America.

This last issue was seen by all politicians, on both sides of the Atlantic, to be the key problem; but like so many matters of bedrock principle which become the inherent fulcrum on which all other motives depend, the question of 'taxation without representation' was not at first raised. This old English Radical cry, which goes back in essence to the Magna Carta, had been raised in the American colonies in 1765, when the British government sought, not unreasonably, to make the Americans help to pay for the war against France, which had cost the English nearly as much as it had benefited the Americans.

There were 3 million people in the colonies, more than in Scotland, Wales or Ireland, and about half as many as in England itself. They were not poor. They were probably much richer than the average Englishman – partly because the ownership of slaves increased apparent wealth, while reducing the numbers of people amongst whom that wealth had to be divided.[12] But Americans, like Englishmen, disliked paying taxes. The stamp tax[13] was withdrawn in 1765, but the following year a resounding declaration was made that the British Parliament had every right to tax the colonies, and trifling, annoying duties ('of principle'), costing far more to collect than they produced in revenue, were imposed on glass, lead, paper and tea. But King George III obstinately persuaded his ministers that, when the other taxes were reduced, the one on tea should remain, though at only 3d per pound. This too would cost more to collect than it would ever produce in revenue, since there was no current likelihood of import-ing more than 5 million lb of tea at, say 1s 6d per pound retail. This would produce about £60,000 in revenue at 3d per pound duty. The actual import of all tea, legal and smuggled (at a much higher price) was only about a quarter of this quantity. The British motives were most curious, to say the least. In May 1769 the government had declared that it did not propose to levy any more taxes on the colonies for the sake of raising revenue; by their own reckoning, therefore, the tea duty was imposed only to establish the right of the government to tax the colonies.

Englishmen were deeply divided on the issue; so, to be fair, were the colonists. But New England was probably overwhelmingly in favour of resistance. After all, for many years the prosperity of the traders and ship-pers of the northeast had relied on an absolutely independent trade bet-ween the French West Indian colonies – flying a successful ribald salute in the direction of the Navigation Acts[14] – and trading in breach of the Indian laws.[15] They made money by smuggling French sugar from the West Indies, turning it into rum, and selling the rum to the Indians; no duty was paid on either the sugar or the rum. Most New Englanders paid scant heed to the government in London. Some Americans had continued to trade with the enemy while their brothers or neighbours or friends were fight-ing the French or the Indian allies of the French. Not for them the modern idea of total war – in stony New England, the sea and its trade represented greater wealth than anything on land, and that trade had to continue, war or no war.

Massachusetts only needed a spark to ignite it when the tea duty came in, and on 16 December 1773 a body of whites, disguised as Mohawk Indians, boarded three ships in Boston harbour and threw the whole cargo into the water. The shores of the tidal reaches of the Charles River were covered with tealeaves. Other parts of the colonies had their own tea parties. Large quantities were destroyed at New York, Greenwich, Philadelphia, Annapolis, Savannah and Charleston. At most of these parties the jokers wore Indian dress. Women of quality met all over to proscribe the more usual kind of tea party. Sometimes they passed resolutions; one from Edenton, North Carolina included these words: ' . . . how zealously and faithfully American ladies follow the laudable example of their husbands . . . ' and they went on to forswear tea in favour of other drinks.

The intellectually respectable reason for American resistance was that the chosen corporation, the East India Company, would have a monopoly of legal tea, sold through chosen merchants who would have a local monopoly, and other trade in other commodities would in turn be awarded by the crown to other monopolists. Legitimate objection was mixed up with the interests of the smugglers of Providence, New York and Philadelphia, until the whole of New England and the middle colonies were yelping about cheap tea interfering with liberty. Tea became not so much a tipple, more a talisman. The party in favour of the English proposal was held to be stuffy, establishment and over-correct in its attitude; the opposition joined the coffee party, which was young at heart, forward-looking and independent in spirit. It is quite extraordinary that this beverage, tea, should have become an important element in the apparently irresistible process of the American Revolution. But since Independence tea has had a slightly un-American feel about it, and loyal Canada has drunk four times as much tea per head than has the independent USA. Tea is identified as the principal non-alcoholic drink of Anglo-Saxons everywhere except in the United States. Can this be another result of Lord North's parlous administration? Some have even claimed that the Boston Tea Party was one of the prime reasons for the Revolution.

The tea parties might have become just a sour joke, but the British reaction was to close the port of Boston, which led directly to the Declaration of Independence. Resistance to tea was a causative factor in the colonists' major struggle against taxation without representation.

The war increased the numbers of Americans who found virtue in Independence, but it was the adhesion of England's jealous European rivals to the American cause which guaranteed the existence of the United States. At the end, the British were fighting their own kith and kin, plus the French, the Spanish and the Dutch, while the Baltic countries formed a neutral but unfriendly Northern League with the aim of preventing the Royal Navy searching neutral ships for contraband. By 1783 the exhausting war was over. England had a more prudent, more careful, more sensible First Lord of the Treasury, William Pitt the Younger.

In 1784 Pitt turned his attention to tea. In that year the duty was nearly as much as wholesale, bonded cost, being 50 per cent of the wholesale price, plus 1s 1d per lb. Thus for the cheapest tea, a Bohea retailing at 5s per pound, the duty was more than 2s 6d, or 50 per cent, and for the most expensive tea, say a Hyson costing as much as £1, the duty amounted to about 8s 6d, or 42½ per cent. As a direct consequence of high duties about half the tea was smuggled in from the Netherlands, or sold at sea by smugglers from the private trade of ships' officers on board the East Indiamen themselves. The revenue was being made ridiculous, the nation divided and disaffected, and the law brought into disrepute. Pitt acted boldly. Tea tax was reduced to a sliding scale of between 2½d and 6½d – about 10 per cent. Smugglers and Dutch tea traders were outraged, but imports of legal tea doubled and everyone else was full of praise for Pitt's obvious commonsense.

In 1801 the English each consumed 2½ lb of tea and 17 lb of sugar (much of it with tea) per head. Sugar 'demanded' slavery: what was the cost of tea? At retail, the nation was paying out £7.5 million; at the English port about half that sum; and about a quarter of that sum, £2 million, in China.

This huge sum of money (equivalent to about a billion dollars a year in today's value) had to be met every year in Canton, and found by the East India Company, the supercargoes who traded for the London importing houses, or by the ships' officers who had the privilege of conducting private trade. The Chinese were not interested in most imports. At that date no new technology existed which could loosen their purse strings. China was self-sufficient in food, textiles, most minerals and all the other necessities of life. For more than two centuries of trade with Europeans, the Chinese, the greatest pre-industrial traders, had accepted only copper, gold and silver in exchange. The enormous rise in the consumption of tea in England, which was paralleled in the Netherlands, Germany, Sweden and the British colonies, could be supplied at that date only from China, and apparently only in exchange for bullion. Concurrently, the French Revolutionary and Napoleonic Wars had placed a great strain upon the finances of all the belligerents: in France so much money was printed (the *assignats*) that inflation resulted; in continental Europe French armies lived off the country, and the French government legalized looting on a vast scale; in England the government abandoned the gold standard,[16] which gave rise to the only serious inflation between 1660 and 1914 in a country where sound money was more important than the views of economists. The world went mad with war, and money lost its value.

Nothing of this kind, however, happened in China. The Chinese merchants, half a world away from Europe, were not interested at that date in paper money. They recognized gold as a store of value, but they preferred silver to gold. Relatively, these true commodities had risen in value against European labour, materials and manufactures by 20 per cent in 1801, and by 50 per cent in 1810. The price of tea also rose, but not by any like multiple. The East India Company, its trading partners and its officers found themselves the classic victims of inflation. Costs rose; product value rose less; result: misery. An answer had to be found.

That answer was opium, which had been an East India Company monopoly in India since 1758, the year after Clive's great victory at Plassey. In 1773 the illicit trade with China was wrested from the Portuguese by the English. China had banned opium in 1729, making its growing, supply or smoking an offence, and ultimately a capital one. The British nevertheless exported 60 tons in 1776 and five times that quantity in 1790; this was all sold to smugglers or to corrupt Chinese. After 1800 the traffic was organized and became an immense industry. The growing and preparation of opium in India was a government monopoly, carefully controlled by the East India Company, which was at that date not only the monopoly trader within India, but also the government. Areas of Bengal most suitable for growing the opium poppy were carefully selected by the English, as advised by Turkish and Persian experience. The white-flowered poppy was grown around Patna, and the red-flowered one in the hills. The collection of the opium juice started about 25 February, and consisted of making the poppy-head 'bleed' its juice daily for two to three weeks. The collection of this juice and its cure, and the preparation of the opium cake by drying, pressing and fermentation, became an industry employing at its height nearly a million men, women and children.

The growing and manufacture of opium at that date not only proved to be from one and a half to three times as profitable per acre as growing wheat or rice, but the added value to the East India Company was of an order of profit which can only be described by the well-known phrase 'the right to print money'. The Indian peasant was, of course, taxed by the Company, so that most of his advantage in growing opium rather than wheat reverted to the Company. The British charged the Chinese merchants about £1500 per ton for opium at a time when tea was about £40 per ton at Canton, and gold less than £4 per ounce. To be historically fair to the British, and to allow that opium required twice the effort in growth, preparation and manufacture of the best tea, and four times that of the average tea, that would still make opium only three times the average value, by weight, of tea. But in fact the ratio over all teas at Canton was nearly 40:1 in favour of opium. In 1830 the British exported nearly 3 million lb, or 150 tons, of opium a year, worth £2 million in the money of those days. This is equivalent to a billion dollars a year in today's money.

The East India Company and the British government rationalized the opium trade with the kind of bland hypocrisy which has made the English establishment a byword for three centuries. There was no direct connection between the opium trade and the East India Company which, of course, had a monopoly position in the British tea trade until 1834 and ruled India until 1858, after the Mutiny. The Company knew of and benefited from the growing of opium in India. The opium was sold at auction in Calcutta. After this, the Company abjured all responsibility for the drug.

The country merchants, that is those who traded between India and other places in the East, bought the opium and took it to Lintin Island in Canton Bay, where it was stored in hulks anchored offshore. Thereafter it was handled by Chinese. Again the Company, whose ships passed Lintin Island on their way to and from Canton, knew quite well what went on, but could plead ignorance to the Chinese authorities. Though the country merchants traded in other goods, notably cotton, opium represented 75 per cent of the trade and was exchanged for silver coin, which was taken back to Calcutta and sold to the East India Company against bankers' drafts in London. Again, this trade was continuous, and the Company could claim that there was nothing unusual in buying silver in Calcutta in the normal way of business; indeed, the Company would always exchange silver for anyone in banking or commerce any day of the year – except Sundays, of course.

The silver then went all the way back to London, the Company taking a cut on the freight and insurance. The Company's agents, or the supercargoes leaving the Port of London, were supplied with the necessary Chinese silver dollars to buy tea in Canton itself. Again, the Company could claim that they did not know that this silver was already dirty money, obtained in exchange for opium. Protests from China to the Company or to the British government were greeted with a shrug and regrets about the inability of the Company or the government to interfere; no wonder the Chinese had a loathing and contempt for the British. An alternative arrangement involved the country merchants selling their cargoes further up the coast, offshore, but in Chinese territorial waters, doing a deal with the opium merchants in junks rather like bootleggers importing alcohol offshore during Prohibition. Efforts were made by the Chinese to check the trade, but they were no more successful than today's Italian or American efforts to check the Mafia. To quote from a Victorian account, *The Fan Kwai at Canton*, by W.C. Hunter:

So perfect a system of bribery existed (with which foreigners had nothing whatever to do) that the business was carried on with ease and regularity. Temporary interruptions occurred, as for instance on the installation of newly arrived magistrates. Then the question of fees arose; but was soon settled unless the newcomer was exorbitant in his demands, or, as the broker would express it, 'too muchee foolo'. In good time, however, it would be arranged satisfactorily, the brokers re-appeared with beaming faces, and peace and immunity reigned in the land. . . .

The Canton officials rarely made any reference to the Lintin station; but sometimes, compelled by force to do so, would issue a proclamation ordering vessels 'loitering at the outer anchorage' either to come into port or sail away to their own countries lest the 'dragons of war' should be opened, and with fiery discharges annihilate all who opposed this, a 'special edict'.

The American tea merchants[17] were not directly involved in the opium-growing, as were the East Indian government officials who controlled the drug culture in Bengal. The American merchants bought opium at Constantinople, Salonika, Smyrna or Beirut, all then part of the Ottoman Turkish Empire. The Turkish government monopoly was less honest than the Indian one, and the opium was adulterated with grape juice thickened with flour, fig paste, liquorice, half-dried apricots, gum tragacanth and sometimes even spirits of lead. After these adulterants had been removed, the Turkish opium was graded as shipping, druggist's or manufacturing quality. The last was used to make heroin, the druggist's opium for the ethical trade, and the very best supplies were reserved for smoking or eating. It was this, the finest grade, which the American tea merchants bought, or had bought for them, and which they shipped direct to Canton or sent via the eastern seaboard of the USA. These merchants were anxious to conceal their vital connection with the drug trade, but the opium element was as important in the American tea trade as in the British. Many a fine family in New York, Boston or Salem grew rich on this commerce, and opium became as much of a permeating problem in the Chinese Empire in the late nineteenth century as is heroin in the USA today. Ironically opium, even of the purest type, is less damaging than heroin, since it takes longer to enter the bloodstream, just as the coca leaf is relatively harmless compared with cocaine injected via the hypodermic needle.

The opium–silver–tea syndrome was a perfect self-enrichment process, yet strangely one not studied by great economists such as Maynard Keynes, who in his youth wrote a book about Indian silver. In order to meet the demand for a mildly addictive drug infusion – tea – which ultimately goes down the drain, the merchants had previously substituted silver, which requires between 1 and 1000 tons of rock to be crushed to produce each ounce of bullion, which is then worth more than most commodities. Silver becomes short and difficult to obtain. Substitute a crop – opium – which is ultra-addictive and goes up in smoke, and the possession of which is illegal at the place of demand. Provided that the supply of the drug is carefully controlled so that it never outruns the increase in the number of addicts, you are effectively 'growing' silver much more cheaply than you can mine it, and so 'printing money', but in a way which guarantees that the recipients of that money destroy it as soon as they can. Significantly, though the poppy can be grown in almost every province of China, all opium was imported (originally only about 3000 lb a year), usually bought by Chinese merchants from the Portuguese. The British increased the trade to 3 million lb or by 1000 times. An expanding demand and an unlimited supply of opium were substituted for a finite supply of bullion. Trade could ask for no more.

The mandarins of China were outraged at the debasement of their people. Opium, probably introduced into China as a painkiller by the Arabs in the eleventh century AD, was used only as a medicine until the eighteenth century. The Chinese government, still at that date not yet completely beyond hope of reform, made effort after effort between 1796 and 1830 to bring the trade to an end, but failed. There were too many addicts, too many pushers, too many 'respectable' merchants making too much money. In 1838 the Emperor Tao-kwang sent a commissioner, Lin Tze-su, to Canton to stop the contraband trade in opium. He issued an order to the Chinese merchants to destroy their stocks and to the British to remove their drugs, but no one paid any attention. The merchants had heard it all before, and had ignored the same kind of orders with impunity in the past. So Commissioner Lin set fire to the Chinese stocks ashore and to British hulks in the harbour. A year's supply of opium went up in smoke in a vast bonfire instead of in thousands of pipes; the smell was said to have been memorable.

The British did not object to this *auto-da-fé*, but continued to smuggle opium ashore at Canton. However, they had misjudged Commissioner Lin, who arrested the British, burnt the opium, jailed British sailors and tortured Chinese merchants. Outraged, the British shelled Canton as a precautionary and punitive measure. Commissioner Lin refused to kowtow. Outrages were committed on both sides. War, at the leisurely pace of these pre-telegraph days, was ultimately declared.

The Chinese in 1840 had no idea of the character, strength or determination of the Europeans in general and of the English in particular. The Portuguese had been limited to Macao since the sixteenth century, and all other Europeans to Canton since the 1690s. The British had carried on a great trade with China for two centuries, restricted to a wharfside less than 800 yards long and 40 yards wide. The Chinese had met and sometimes defeated European ships at sea; more often, they had lost. The Emperor and his advisers probably did not hear of these failures. No reports were carried back to landlocked Peking about the failures of Chinese pirates and naval vessels at sea in the far south, in the China Sea itself. Dead men tell no tales, and the vanquished found it difficult to make rulers accept that there was anything else to explain defeat other than their incompetence. Europeans were still regarded as ignorant foreign devils, barbarians, only interested in money, drink, trade and women.

In 1834 the East India Company monopoly of the China trade, and of course of the British trade in tea, came to an end. New men, enlightened, competitive free traders, nineteenth-century liberals to a man, succeeded the comfortable and corrupt servants of the Company. The pressures on the home government to 'open' China increased. The outrages at Canton, of which the British complained so much and which formed the *casus belli* of the Opium War of 1840–2, were probably as much a Chinese response to the changed circumstances as to the long-term debauchery of opium itself. To the Chinese mind, then as now, if you have to deal with foreign devils it is better to deal with a fixed group of men whose character and strength and weakness may be gauged, than to have to assess an ever-changing cast of characters in a long-running play called *The Market Economy*. The Chinese knew the Cantonese Europeans of old. They did not know the new men.

But the new men were determined that China and Chinese trade of all kinds should be opened to the West. The British took the initiative, secure in the power and pre-eminence of the Royal Navy. The Chinese did not have a chance. The first part of the war was short and decisive. In 1840 Chusan was captured, and the following year the British bombarded and destroyed the Bogue Forts on the Canton River. The local Chinese commander, Ki Shen, who had succeeded Commissioner Lin, agreed to cede Hong Kong and pay an indemnity of 6 million Chinese silver dollars, worth about £300,000 then and about $20 million today. When the news reached Peking, the Emperor was persuaded that Ki Shen was incompetent and unpatriotic. He was degraded and exiled to the country. War was resumed. Inevitably, the Chinese lost. In the aftermath Amoy, Fuchow, Ningpo and Shanghai became 'open ports', while the government had to pay a further indemnity of 21 million silver dollars to the British and to accept European supervision of the Chinese customs. No mention was made of the import of opium.

For a further sixty years the trade in Chinese tea exceeded the trade in tea from all other countries. In fact, in the hinge year of 1840 the first puny export of tea from India and Java had begun: less than a ton was sold in London. Chinese exports to all countries in the previous decade had averaged over 200 million lb (100,000 short tons), worth about £25–30 million in contemporary terms, or more than $3 billion in today's money.

Throughout the rest of the nineteenth century China was gradually opened up, aided by a succession of wars in 1856, 1861, 1871 and 1894. The period was characterized by the gradual decay of the central government in Peking, occasionally delayed by the employment of vigorous, sometimes foreign, generals including Charles Gordon. Russia pursued a deliberate – sometimes open, sometimes covert – policy of imperialism in the north. The British and Americans were primarily interested in trade and the making of money, apart from their illusions that the everyday Chinese was one of nature's Protestants. The French had the same illusion about Catholicism and the same greed about trade. The rivalry between merchants was almost exactly mirrored by the rivalry between Christian sects. The Chinese did not reveal their preferences, and acquired the reputation of being inscrutable, though perhaps they were only resentful and confused. The opium trade accelerated, imports almost exactly matching the deficit in the balance of payments which the West had with China. Opium represented one-sixth by value of the imports into China during the nineteenth century.

The material conditions of the Chinese did not improve. Spiritually, despite the missionaries – who were to form such a massive lobby in the next century – the huge country was in decay. Politically, the once efficient dictatorship gave way to local despots called T'ai-pings, who were usually corrupt and nearly always a more onerous alternative to central government. The material treasures of China were destroyed or dispersed all over the world. The loss of this cornucopia of two millennia of civilization was matched by the destruction of the Chinese genius for craftsmanship and design. The Chinese became copiers and coolies, hewers of wood and drawers of water for the West.

The wrongs committed against a relatively weak China have been as damaging as any transgression against other non-European peoples. Perhaps the insidious nature of the opium evil should be noted and the advanced state of Chinese technology accepted. From iron- and steel-making, to pumps, to mills of all kinds, to canals, irrigation and other water management, textile machinery, harness, crossbows, concave ploughs, bridges of all three types, sternpost rudders and watertight compartments in ships, fore-and-aft sailing rigs, magnetism and compasses and gimbals, paper of all sorts, as well as gunpowder and porcelain, China was between four and seventeen centuries ahead of all European nations (see table on p. 96). We forget this, in our Eurocentric way – most people know only about paper, porcelain and gunpowder. And this materialistic list takes no account of Chinese supremacy in astronomy, biology, medicine or any other area.

China, a repository of arts and artifacts, of craftsmanship, design, ingenuity and philosophy, was raped for a few years' increase in the national income of the Europeans. For a pot of tea, one could say, Chinese culture was very nearly destroyed. Whether it will ever recover under the post-1950 regime is another matter, which has nothing to do with tea.

After the East India Company's monopoly in tea came to an end, the first ship to arrive with a 'free' crop of tea was called the *Carnatic*; she is celebrated in the name of the road in Mossley Hill, Liverpool, where her owner lived. The East India Company knew that they could not compete and sold their entire fleet of dumpy, heavy ships, each containing up to 1000 tons of well-seasoned timber. The ships were so well built that one of them stayed afloat until 1897. Another was sold for breaking-up, and the timber was so valuable that the ship made over £7500 (more than $3 million in today's money).

The 'free' tea trade developed in Liverpool, Dublin and other ports for the first time, since all East India Company ships had previously landed their cargoes in London. There was great growth in tea imports, and a great growth in the instability of the market, since no single organization, Chinese or British, continued to hold stocks as had the East India Company and its Chinese trading partners in Canton. The Opium War not only interrupted trade for nearly three years, but the disruption in China probably destroyed or damaged as much as one year's complete stock of tea. In the 1840s no one, anywhere in the world, carried large buffer stocks.

The East Indiamen had travelled in a most leisurely, comfortable fashion, outward via the Cape of Good Hope, which was British after 1815; east, sometimes, all round Australia and then due north to China, avoiding the East Indies altogether. The squat, matronly East Indiamen, 'floating warehouses' built for comfort rather than speed, sometimes anchored at night in fog or when navigation was difficult and had no incentive to sail at speed, because no one apart from the Company traded in tea. After 1815, however, the Pax Britannica, maintained by the Royal Navy, meant that self-defence and long voyages avoiding land were no longer necessary. The safe, slow East Indiamen gave way to competitive ships, trading for a product whose supply might well be short, and whose sale was enhanced by early disposal. Finer ships with a length/beam ratio of 5:1 or 6:1 replaced the Company ships whose ratio had been 2½:1 or 3:1. During the 1840s British sailing ships gradually became sleeker and carried smaller crews, the combination making for higher profits but less safety. In 1850 the Navigation Acts were repealed, and the ships of any nation were free to enter British ports with any cargo; previously, foreign ships had been limited to cargoes from their own country or Britain, and were debarred from carrying 'third nation' freight. But only the Americans were ready and willing to compete in the China trade.

During the 1850s the short reign of the China clippers began. The clipper ship was built for speed and the carriage of valuable cargo. The length/beam ratio was increased to about 8:1; the bow was flared and sharpened to slice through the water rather than bash and shake and slow the ship in a seaway; the foremast was brought aft by about a tenth of the ship's length, to prevent the foresails driving the bow under water; the masts were raked to increase the effective sail area; the stern developed from a massive wall into the graceful 'clipper stern', still used in steamships after World War I.

Clippers were not the first fast ships; there had been many fast, small frigates, revenue cutters and smugglers' sloops. But the clippers were the first fast big ships, and set up all sorts of problems for naval architects, ship-owners and the men who sailed in them. The risk of losing ship and cargo probably doubled, but the revenue-earning capacity of the ship probably also doubled. A ship now took 90–120 days to sail from China to New York or London, instead of the 180–270 days of the East India Company fleet. New-season tea arrived much more quickly. Clippers had been developed in the USA to take high-value cargoes at speed: illegal slaves; mail; rich Atlantic passengers; emigrants to California round the Horn. Clippers were used in every one of the Seven Seas, but their fame rests on the tea race.

The economic justification for the race is difficult to assess. Properly packed, tea does not deteriorate. In the old days it had sometimes taken a year to assemble a cargo in Canton, and nearly a year to reach London, followed by another year in store. Carefully handled, tea nearly three years old does not taste any different from tea three months old. (In 1955 a tea merchant made and drank some tea that he had found in a sealed cask belonging to his great-grandfather, who had died in the Opium War in 1840; 115 years later it was still perfect, delicate, and unlike any modern tea in its subtlety.)

The public, therefore, was somehow persuaded that 'new-season' tea was in some way different from last year's tea. It is easy to see the advantage to the merchants. If the public is brainwashed into the belief that tea goes 'off', the merchant can clear his stocks every year and sharpen his turnover. For the public, the advantage is less obvious. But no one stopped to think, and the clipper races became as exciting as any other race. In the 1850s Americans and British raced impartially from China to New York or London, since there was not much difference in the distance, respectively, round the Horn or round the Cape of Good Hope. In the 1860s the Americans were otherwise occupied, first in the Civil War, and then in the trade with California, Alaska, Japan and the east coast. After 1862 the clipper races were exclusively British. When the Suez Canal opened in 1869 the races ended and the expensive sailing ships had to find other employment, carrying emigrants to Australia, New Zealand and the American west coast.

Clippers charged £5–6 per ton for tea, China–London, double the slow-ship rate, and with a big bonus if they got there first. Snobs would make a point of saying that the tea came from the clipper *Ariel*, or *Era*, or *Cutty Sark*. This is similar to those people who will pay a premium price for 'new season's' grouse on the evening of 12 August, in an age when on 11 August anyone can take out of the freezer a grouse which, cooked properly, will taste the same as the bird a year younger. The clipper races were supported by atavistic, deep-seated nonsense, and they were finally killed by steam.

Early, inefficient steamships with low-pressure boilers had to carry so much coal, even if helped along by sails, that they could only just cross the 3000 miles of Atlantic without refuelling. To reach China round the Cape of Good Hope they had to refuel at least once in the Atlantic, once at Cape Town, at least once in the Indian Ocean, and again in Singapore, and there is little evidence that any steamship line made much profit before the Suez Canal was opened. After 1869 steamships could beat sailing vessels, which still had to go round the Cape rather than through the Canal since the winds in the Red Sea were too unreliable. For bulk cargoes, where time was of no importance, the Cape continued to be used until the 1880s, when the more efficient steam engines that had been developed finally started to beat sail for all cargoes other than those of low value. But for many years it was only passengers, mail, silk, tea and other light, important freight which went through the Canal. The steamships needed the Canal, but the Canal needed the steamships just as badly. The symbiosis would not have worked a decade earlier, and a decade later steamship efficiency might have increased so much as to render the Canal unnecessary.

Suez and steam together made one great difference to all high-value cargoes. Stocks were no longer maintained in a long, slow pipeline to and from the East. They were held at either end, separated by less than fifty days' voyage time, half the time of the fastest, most dangerous, riskiest, most romantic clipper which needed a picked crew, a high freight rate, and a great deal of luck. Steam, smelly steam, had brought the tea trade into the industrial age.

Early in 1820 David Scott, commissioner for the newly acquired state of

Assam in British India, sent samples of leaves from Cooch-Bihar and Ran-pur to his superiors in Calcutta. Here they were declared to be leaves of one of the innumerable species of camellia, and sent by Dr Wallich, the government botanist, to London where they were re-examined by the herbalist of the Linnean Society, and pronounced to be leaves from the tea plant. This was the first known discovery of a wild tea plant in India, which at that date had no tea plantations. At this time nearly all tea came from China, except for a small export surplus available in Japan, and even less in Formosa.

Ironically, when the tea industry started in Assam a dozen years later, the tea-gardens were planted with cuttings from Chinese trees, which died, or did not thrive, or failed to become productive. The native Assamese wild plants had been uprooted and burnt to make space for them, and now the hills were scavenged for more samples of the wild plant from which to breed better-suited specimens. Yet the native Assamese people had never, as far as the British knew, used tea to make an infusion. This was conventional wisdom, though in one or two places on the banks of the Brahmaputra River there is evidence in abandoned gardens of ancient cultivation of tea. But there is no native Assamese folklore about these plantations.

Tea is much more than a garden crop harvested to make the raw material of a drink. It is complicated by the methods of drying the wet material, which may ultimately turn out green or black. Green tea is made from leaves dried shortly after picking so that the chlorophyll, and much else besides, is not subject to great changes. Black tea does not come from any special place, but is the green product subjected to a process wrongly called 'fermentation'.

The young leaves and buds of the tea plant may be pickled and made into 'Leppet tea', as in the Shan States in the mountains between Burma and China, and eaten as a vegetable. Green or black tea may be made into cakes, which in Tibet are turned into a thickish soup and, mixed with rancid yak butter, slurped or eaten with a spoon – an acquired taste. Tea may be chewed, either green or black, as in some parts of Indo-China. It may even be ground and sniffed like snuff, as in Yunnan.

Brick tea was made for export to Russia and Mongolia, because it was relatively easy to transport in this form by camel or horse caravan overland from China, taking six months to get to European Russia. The tea was of the highest quality, the best dust and siftings, steamed and compessed into bricks under hydraulic pressure. A great brick tea industry started more than three hundred years ago, becoming mature in the 1860s, ultimately involving Chinese labour, Russian supervision and French and British capital. Brick tea needs a kettle or samovar in which to brew it, and it is notable that this type of tea was so popular that the industry continued long after steamships rendered it unnecessary.

Perhaps 90 per cent of all the tea sold in the last hundred years has been black. It has been sold, outside the countries of production, in packets to the ultimate consumer. Adulteration has become rare, not because of a change for the better in human nature, but because the practice ceased to pay.

Tea has spread to many countries. It has been grown in its original home, China, and in Japan, Formosa (Taiwan), Burma, India, Malaysia, Ceylon (Sri Lanka), Indonesia, Iran, Turkey, the Philippines and Queensland, as well as in a dozen parts of Africa, Indian Ocean islands and Latin America; South Carolina, Argentina and Georgia in the USSR are the extremes in geographical terms and the furthest from the original centres of cultivation. The absolute necessities of tea cultivation are climate, which must be wet and warm; soil, which should be deep, friable, high in humus and have a pH[18] of about 5.0 to 5.5; and labour, whch must be cheap and plentiful. These are the conditions which led to the development of the European tea industry in India, Sri Lanka, Indonesia and Africa, in order of original successful plant transfer and in the same order of tonnage of tea today – more than 80 per cent of the world trade in tea.

The process of plant transfer was a hit or miss affair. After China was opened up following the Opium War, the botanist Robert Fortune, a Scotsman from Berwickshire, finally corrected the Linnean misclassification[19] by observation in China itself in 1842–3. This story is important, not because it denigrates Linnaeus but because the European tea industry, and specifically the British tea industry, had grown into a vast affair with an annual turnover of millions of pounds sterling, and had made the East India Company the IBM or General Motors of its day – and it was all done through trading in a product whose origins were unknown for two and a half centuries. After much experiment and many failures, an industrialized European tea industry had been successfully established in Assam in about 1860, and in Ceylon and Java by 1890.

The Assam tea plant is of an agrotype, or ecotype, which, if left to its own devices, would reach a straggly height of 30–60 feet. In cultivation, however, it is plucked and pruned into a convenient, 3-feet-high bush which may be reached with ease by the picker. There is a correlation in the agrotypes between response to fertilizer, leaf type, morphology of the cell structure, and yield of raw tea per acre. Selection has therefore become intensely practical and discriminating. In Japan, where the highest yields are obtained, an average of more than 1½ tons of finished product per acre has been achieved; in some gardens, 2 tons; in one recorded case, 3 tons. The wrong type of bush, grown in the wrong way, will not yield one-hundredth of that quantity of tea.

All this kind of husbandry took many years to develop. Quality took second place to quantity, and for a long time, at least until after World War I, teas from India and Ceylon had a reputation for being second-rate. This reputation was perhaps undeserved, but it is also true that, if China had not been destabilized and subsequently destroyed by Western penetration, then India and Ceylon would not have become tea-producers. Despite all the political troubles which finally destroyed the Empire, Chinese tea exports were higher than those of the whole of the rest of the world until 1890, and higher than any other single country as late as 1910. During this time Indian tea-planters were still trying to imitate the Chinese product. In the end, it was realized that manufacture begins in the field, and that however much effort is made to mechanize the curing of tea, it is the growing that remains a precondition of a successful, saleable product. If the correct ecotype is planted, pruned and fertilized, and the picking matches the season, the desired result will be much easier to obtain. Manufacture is carried out on the estate, and the post-picking process was an early candidate for mechanization. Picking itself is another matter.

The bushes are picked more frequently if a higher quality is required, and less frequently if greater quantity is the aim. The first spring pick makes the best tea. Frequency of picking depends on weather, fertilization of the soil, pruning and so forth. The tealeaves should be picked dry, but this is not always possible, and large quantities of low-grade material result from 'rains teas'. The industry might be better off if these inferior grades were thrown on to the compost heap.

The picked leaves can be turned into three different products, though each estate tries to concentrate on one speciality. Black tea represents over 95 per cent of world trade today; green tea is of importance in the Far East and a cult in the West; Oolong is important in China and Taiwan, with a small export to the USA. Black tea is fully processed. Green tea is fired before any oxidization has taken place. Oolong is partly 'fermented' and then fired early. All living vegetable materials are covered with organisms which permit the recycling of the material as soon as 'death' occurs, so when tea leaves are picked the enzymes start at once to recycle the green material. In order to control this process, the green leaves are allowed to wither naturally by being spread out on special racks. Here they lose, by evaporation, 50 per cent of their moisture. If the ambient humidity is naturally high, as during the rains, the process takes longer, but the aim is to wither within a maximum of 12–16 hours. A longer wither is one more reason why 'rains teas' are inferior.

After withering, the teas are rolled by machine in order to accelerate oxidization, and this process, an essential precursor to 'fermentation', was formerly done by hand. Of all the mechanical innovations in the industry, the rolling machine pays the highest dividend. One machine today can do the work of up to a hundred people. The rolling goes on until the still-green tea is a mass of pulp. Strict control of the severity and timing of the rolling is necessary to prevent the green, partially dried tea turning into a pulpy mass resembling cold, overcooked spinach.

In the case of black and Oolong tea the so-called 'fermentation' follows. This is done in another special area, in a temperature of about 80–82°F, and the green mass turns coppery red in colour. The temperature, depth of mass and evenness of oxidizing action are all necessary elements in achieving quality. The most critical decision, however, and difficult to make other than by instinct and experience, concerns the right time to terminate the process by moving the mass into the firing cylinders. This choice is dependent upon the judgement of eye and nose, and is often varied from batch to batch according to local conditions.

Firing involves virtually the same process as does high-quality grass-drying. The moisture content of the withered, fermented mass of tea has to be reduced from about 45 per cent to about 5 per cent, and this must be done without affecting the quality of the tea. Too high a temperature produces a taste nearer to burnt toast than to tea. Underfiring leaves the infusion more like green tea than black, even if the tea is black in appearance. Firing can be mechanized, but human control must be continuous, skilled and devoted.

After firing, the tea must be sorted. The names given to the various teas are so similar that they are very confusing. The main grades are broken and small leaves, which produce the grades (not the quality) called broken orange Pekoe, flowery orange Pekoe, broken Pekoe, Souchong and so forth. These are all 'small' teas, and the two smallest grades are called fannings and dust. These are not, as their names suggest, sweepings, but the smallest, youngest and probably 'strongest' teas. They were formerly the constituents of brick teas, but nowadays they nearly all go into teabags. The bigger teas are called flowery orange Pekoe, orange Pekoe and Pekoe. The word 'Pekoe' appears in all the grades except fannings and dust; the word 'orange' appears in more than half the grades, but has nothing to do with taste or oranges; the word 'Souchong' appears once, but has no meaning as in the phrase 'Lapsang Souchong'. The tea trade does itself a great disservice by this obfuscation, especially as these expressions mean different things in different states to the producer and in different countries to the consumer.

Though the (mostly female) picking force has proved impossible to replace by machine, efforts were made in the hopeful 1920s in Soviet Georgia to use a modified hedgecutter for the job. Needless to say, it was a great failure. The Japanese invented a form of shears with a bag suspended below, which looked disconcertingly like a pelican's beak and which was also not a success. Both failures were the result of inability to discriminate. Two leaves and a bud are the aim of the clever fingers of the underpaid human picker. No machine short of a laser-controlled robot could approach a skilled woman in speed, accuracy or reliability – as long as the supply of labour continues. One day, perhaps, the picking problem might be solved. If the process were mechanized, a great change could come over the industry. At least six American states would be able to produce tea of a high quality, as was once achieved in South Carolina. Only the high cost of picking prevents such a development today.

The other three candidates for mechanization are withering to 50 per cent moisture; rolling; and drying to 5 per cent moisture. Withering by the use of heated air gives a poorer-quality product, so that during the rainy season air is pre-dried by being passed over silica gel, thus achieving the same result as with hot air but without added temperature. This was a development of the 1930s, but as early as 1850 withering fans were in use in India, and from the 1870s in Java. A huge machine like a domestic tumble drier was developed to finish off the wither at 100°F. No machine except the silica gel drier does as good a job as a cool loft in the dry season.

Rolling is an easier problem to surmount. Machines were in use in the late 1880s, and little improvement has been needed or made since. These machines effectively release all the juices from the withered leaf, and they destroy the leaf surface so that air can enter, oxidize and 'ferment'. During the rolling process, which may take up to two hours per batch, the temperature of the mass rises, and efforts have been made to restrict this temperature rise by cooling the rollers with water. Technology has been borrowed from the tobacco, meat-processing and dairy industries. Since 1870 over five thousand English patents have been filed on this subject alone. Few have been manufactured even as prototypes, let alone entered quantity production.

Between 1870 and 1890 no fewer than two hundred successful driers were invented. Early models resembled Roman corn driers, while the later types lent their technology to the whole of twentieth-century agriculture. Many of today's grain, grass or seed driers are based on the tea industry's efforts in the period before the rest of agriculture was out of the horse reaper stage of harvesting.

One manufacturing method which was developed in the 1930s was CTC or 'cut-tear-curl'.[20] Like manufacturing conveniences in other industries, CTC made the achievement of a standard product very much easier and surer, but while it eliminated the lowest quality it also destroyed the best. It is a form of manufactured standardization. CTC makes average dry tea out of almost any leaf, and may well save rains-picked teas. The machines make all teas more highly oxidized, and produce an infusion which is harsher, more highly coloured and brisker than the same leaf more traditionally cured.

Since 1960, an ever larger proportion of the world's tea crop has been marketed in Europe in teabags by supermarkets and huge importers whose product has little identity and is based on fannings and dust. For this kind of mass production the CTC machine is ideal, and the natural alliance between CTC and the teabag is disastrous to quality. Teabags of the 'pillow' type, with a surface area several times that of the traditional teabag, make better tea, but cost, in packaging, about five times as much as the cheapest heat-sealed bags.

The other modern development, instant tea, is a complete failure. No process has produced a drink which approximates the 'real thing' as closely as do the better instant coffees. In fact instant tea bears no relation at all to proper pot-made tea.

'A good cup of tea' is becoming more and more difficult to find unless it is made by a devotee at home. Picking apart, the raw material is handled like dried grass or alfalfa, bundled into huge machines, reduced to a blended dust, and then packed in little bags whose paper imparts taste and whose contents only produce an infusion with coarse strength and lots of colour.

Subtle tea is as far from the bag as is a bottle of plonk from a château-bottled claret. It requires five to seven minutes for infusion; the correct leaf for the local water; and a certain degree of ceremony. Apart from any other factor, tea requires freshly drawn water for each brew: unlike coffee, tea absorbs oxygen, and therefore cannot be made with water from a simmering kettle. As a quick, non-alcoholic 'fix' such a product can never compete, in convenience terms, with instant coffee.

So tea must necessarily lose the battle for convenience, on the grounds of speed and time. There remains the question of quality, and this can only be guaranteed if the tea-drinker also closely supervises the brew itself.

There is one other point to be made. In 1840 no native Indian ever drank tea, except as a Europeanized foible. Today two-thirds of Indian production, the greatest in the world, is consumed domestically. The Indian government's interest lies in cheap, widely available and not very interesting tea for its people at home. The discriminating tea-drinker is fighting a losing battle.

Chinese civilization, then, was debased and almost destroyed by tea and opium. In the light of this it is interesting to consider Japan, a country whose history might have evolved along similar lines, but, as it turned out, did not.[21]

Today the Japanese tea industry is in agronomic terms perhaps the most efficient in the world. Given Japan's well-known discipline and genius for organization, this was probably true in the seventeenth century. Yet in the critical development of the tea trade with the West the Japanese took no part, for from the 1640s until the mid-nineteenth century, the period between the start of the tea trade and its near maturity, Japan was almost cut off from Europe and Europeans. But Japan has not always been isolated. For eleven centuries the country was proudly secluded. Then, for a hundred years from 1541 to 1641, the Japanese welcomed trade to a greater or lesser degree with merchants from Portugal, Spain, the Netherlands and England. Japanese ports were opened to the commerce of the world. They wanted silk and silver; they exported gold and copper. The Europeans also acted as middlemen between Japan and China, and between Japan and European trading stations in the East Indies, India and Africa. From about 1550 until 1616 the Japanese traded with Europeans at home, in Osaka, Nagasaki and Yokohama, and all over the East. The Japanese bargained and exchanged with the Spanish in the Philippines, with the Portuguese in the Indies, and with the Dutch in Formosa. The Dutch in particular were of great importance as intermediaries with China, who would not countenance direct trade with the foreign devils, the Japanese.

In the mid-sixteenth century the Japanese were known as the 'Kings of the Sea'; they had perfected a form of sea warfare based on the clever manoeuvring of ships, which defeated the Koreans and dismayed the Chinese until that inventive people devised a tactical antidote. On land, Japan had mounted and sustained an invasion of Korea which involved an army of 200,000 men and 500 ships, forces which few European powers were to launch for two centuries or more.

The Japanese homeland was a federation of loosely connected feudal satraps, nominally under an emperor in Kyoto, but some of the greater feudal lords were as independent as any duke of Burgundy in medieval France. During this period three far-sighted superlords, Oda, Toyotomi and Tokugawa, had acted in sequence to unite the country so that it became a cohesive state under the emperor, rather than a confused and warring country – feudalism at its most chaotic. The foreigners were gradually squeezed out of the right to travel freely all over the country, and they were confined to certain factory ports, as were Europeans later in China. In the end, all white men except the Dutch were expelled. This happened in 1641, and for two centuries after that the Japanese were in a position of self-imposed isolation.

The Dutch were confined to one island near Nagasaki, a mere strip of sand and shingle 200 yards long and 3 yards wide. While in harbour their ships had to take guns, rudder and sails ashore, and offload their ammunition. Thus rendered safe and immobile, the ships were unloaded at Dutch expense, and the trade conducted on Japanese terms. No Japanese was allowed to speak to any foreigner unless another was present to note what was said. The Dutch were not allowed to be buried ashore, or to go ashore from the island to the town of Nagasaki, or to entertain anyone in a Dutch house or ship, except for 'public women' (the Japanese were always practical people).

These severities were imposed upon the Dutch, and all other foreigners were excluded, for one reason only. For the fifty years up to 1640 Japan had suffered Christian missionaries – mostly Roman Catholics, but some Protestants. In addition to proselytizing large numbers of Japanese, these missionaries also took up the secular cause of their homelands. Spanish Jesuits intrigued against the Portuguese, and both Catholic sets of nationals schemed against the Protestant Dutch and English merchants. Elizabeth's war with Spain, Spain's war with the Netherlands, and all Europe's Thirty Years' War were mirrored in Japan, itself a victim of a complicated feudal struggle.

By 1641 the Japanese had had enough. The exclusion of foreigners was not mere xenophobia: it was specifically anti-Christian, because it appeared that the Japanese could not trade with the Europeans except with the implicit toleration of Christian missionaries. By the early seventeenth century the Japanese position was simple: they wanted trade, but if trade involved missionaries, they would do without all foreign contact. Even Japanese ships were denied the right to go abroad, to leave Japanese territorial waters. In 1638 the Dutch had put the date (according to the Christian era) upon their new warehouse in Osaka. A mob, incensed, or bribed, or otherwise induced, collected to challenge this presumption. The Dutch trading chief, one Caron, saw the mob approaching and without a moment's hesitation set four hundred of his own men to tear down the offending warehouse. Cheated of its prey, the mob dispersed and the Dutch were saved. But they were removed to the tiny island of Deshima near Nagasaki, where they survived on a diminishing trade for 217 years. They had to agree not to pray in public, not to hold services on land or sea, and not to celebrate the Sabbath. This sacrifice of Christian principle led the Dutch to survive as traders in Japan; all other Europeans, traders as well as priests, were expelled and Japanese Christians were executed or forcibly reconverted. The Europeans objected to maltreatment of Christians, but it was no worse and not much better than that inflicted by Protestants and Catholics on each other in Germany at the same time. So Japan became self-sufficient except for a few Dutch ships (ultimately only one) a year, carrying the essential silk from China to be exchanged for the equally essential copper from Japan.[22] The nominal reason for the expulsion of all foreigners, then, was Japanese distaste for Christianity, but there is another possible reason for Japan's vehement hatred of foreigners which had nothing to do with religion.

The whites brought to Japan not only Christianity, but also syphilis from the Americas, via either Europe or the Philippines; the potato from Mexico, also via the Philippines; tobacco from the Americas, via the Netherlands; and gunpowder. Syphilis was controlled by scrupulous hygiene, and the potato was gratefully adopted. Tobacco was outlawed, though not for medical reasons: most Japanese houses were built of wood, paper and textiles to avoid major earthquake damage, and were thus very inflammable, and smoking in bed had been a cause of serious fires in Japanese cities before 1620.

No native European material product was introduced, but, as mentioned earlier, traders did bring gunpowder, which had been discovered in China and had moved to Europe to become a weapon of war. The Chinese (and probably the Japanese) knew about gunpowder as a propellant and explosive for wartime use four hundred years before the backward Europeans. It is entirely possible that China, as well as Japan, had rejected its use for aggressive purposes because it rendered the whole pyramidal structure of feudalism too shaky to survive. Did gunpowder help destroy feudal order in Europe? Almost certainly. Did the Orient know about it? Surely not. But the oriental nations were logical in a way that the hasty, individualist white man would never be. If guns threatened the good order and discipline of feudal society, then guns and the Europeans who brought them would be abandoned to save feudalism.[23]

An end to Japanese aloofness came not from Europeans, now involved in trades of all sorts in the Orient, but from the American North Pacific whaling fleet. In 1823–4, eighty-six American whaling ships passed within sight of Japan's most northerly island, Yezo. American whaling schooners were shipwrecked from time to time and the survivors sent to Batavia in the Dutch East Indies by the single Dutch ship allowed to trade with Japan. Japanese fishermen and sailors turned up in California or Oregon in the 1840s, driven 6000 miles across the Pacific by bad weather. Commodore Biddle was sent to begin trade and consular relationships in 1846, but was politely and positively told to leave, which he did, without setting foot ashore. Various European nations made further approaches, using both the carrot and the stick as inducements for the Japanese to open up their country to the West. Finally Commodore Perry, USN, brought more than two hundred years of seclusion to an end in 1853.[24]

This isolation, unmatched for obvious reasons in the Eurasian landmass, led to a prolongation of feudalism and a self-sufficient autarchy with an extraordinarily narrow but highly developed form of civilization.

There was no knowledge of the universe, of the theory of gravity, of the differential calculus, of the circulation of the blood or of electricity. But in the eighteenth century Japanese agriculture was the most highly developed in the world, as was fishing and fish-farming. Botany was ahead of anything in Europe. On the other hand the world of medicine was limited to herbal remedies, little use being made of surgery or compound drugs. Mathematics was unknown in terms of algebra or geometry. Astrology was known, but not astronomy, and the Japanese could not construct a calendar without Dutch or Chinese help. They disputed with the Chinese the invention of printing, but the art was more advanced than in Europe and they could print in full colour. They were much addicted to poetry, music and painting, and their textiles and ceramics were of a higher standard than those of Europe, if not comparable to the best in China. They were convinced of their natural superiority, and of the unnecessary nature of foreign intercourse. This led to three characteristics which enhance the Japanese character today and are often baffling to the Westerner.

The first is that because Japan was a homogeneous, integral civilization when the white man arrived in force for the second time, in the nineteenth century, the Japanese were able to absorb the invasion of ideas without their society being destroyed, as happened in China, America and Africa. They have adopted the apparently desirable and rejected the apparently unsuitable without detriment to the form and nature of their own complicated culture.

Secondly, for centuries with only one short break the Japanese have had to be self-sufficient. They have had to count their blessings and live with what they had, refusing to become addicted to any foreign import. Their staple foods were rice, fish and the long radish; they did not crave lamb or beef. Their drink was tea or rice wine; they did not crave beer. Whatever they did, they did well. Their land was the most productive in the world, before artificial fertilizers came along, and every farm was treated as a garden. Even today their tea-gardens, their rice paddies and their vegetable patches are the most productive per acre in the world. Their industry was equally efficient. They produced the best steel in the world before Bessemer, the best ship-building technology in wood, and the best answer to building in an earthquake zone before ferro-concrete. In World War II their self-sufficiency and ingenuity allowed them to defeat the British in Malaya with a much smaller force, mounted on bicycles; to produce the most economically efficient fighter aircraft of the time; and to survive in a vast theatre of war against the most powerful nation on earth. They were defeated not by technology, but by science expressed as the atomic bomb, and power, as in the monstrous increase in energy employed by the Americans.[25]

There is a third point. The Japanese were cut off from the world for the most important period in the history of science. Within the limitations of their lives their technology was very efficient and appeared to answer problems. This respect for technology, the ability to make other people's ideas work in their own context, this infertility of invention, has led the Japanese to be called copyists in the first half of the twentieth century and to be feared in the second half as the industrial power which can turn ideas into hardware faster, more economically and more profitably than anyone else, whether with the robot, the laser or the chip. The Japanese are the finest technologists in the world today, and they think practically, technologically, not scientifically. They answer the question 'How?', very rarely the question 'Why?'

Whether Japanese isolation from 1641 to 1853 was caused by religious (not racist) xenophobia or by their desire to do without gunpowder, the effect has been more than profound. The decision taken in 1641 has made Japan the only non-white country to have resisted Europe. This is what, in different circumstances, could have happened in India or China. Japanese history is a tribute to isolation from Europe, the result of not being in the tea trade.

Notes

1   Quoted in *Encyclopedia Britannica*, 3rd edition, 1788–97, under Tea.
2   Charlady, charperson, to char: derived from an archaic word 'char', meaning 'work', and used as a noun, a verb and a prefix.
3   See Noel Perrin, *Giving up the Gun: Japan's Reversion to the Sword 1543–1879*, Random House, 1980. The purity of the narrative in this book, most of which appeared first in the *New Yorker*, is somewhat modified by a secondary argument that what was possible in one feudal country should be viable in the whole world: i.e. Japan gave up the gun, therefore the world can give up nuclear weapons.
4   The Jesuits – the Society of Jesus, founded in 1539 in Spain by (St) Ignatius Loyola – had spread all over the then known world within a generation. They were educators, propagandists and, in Europe, warriors of the Counter-Reformation. In the non-European world they established a missionary standard which few could later achieve and none excel. They arrived in China about 1570, and were much more acceptable there than in Japan, which provided the Society with saints and martyrs from about 1600 onwards.
5   With reasonable winds a sailing ship averaged between 100 and 200 miles per day, but this might be the equivalent of only 60–80 miles on a direct course – the vessel had to allow for contrary winds and tides and currents. South America might thus be nearly six months from Spain, North America nearly three months from France or England, and the Caribbean somewhere in between. A letter sent from the East Indies could not expect a reply in less than a year.
6   The whole complicated spice trade, which was the great motivating factor in eastward exploration, was obstructed by the Turks in the later fifteenth century; it

then became a temporary Portuguese monopoly, which was in turn challenged by the Genoese and Florentines, and was then subject, as it had originally been, to Venetian dominance, if not monopoly. Meanwhile, by 1600 other objects of trade had been discovered in the East, and had become more important than spices.

7 Probably after John Bull.

8 The minimum distance from the English Channel to, say, Canton, via the Cape of Good Hope, is less than 12,000 miles. This is increased in a sailing ship, which is incapable of making a straight-line course, to about double.

9 In his *Genera Plantarum*, published in 1753, Linnaeus included all teas under the name *Thea sinensis*, and recognized two camellias, *Camellia japonica* and *C. sassanqua*. The camellia was named in honour of a Moravian Jesuit called Came (Latinized as Camellus), who lived from 1660 to 1706 and wrote about plants in Asia. In 1762 Linnaeus distinguished two varieties of tea, *Thea viridis* and *T. bohea*. The former means 'green'; the latter is the name of a range of mountains from where, it was once thought, black tea came.

For two centuries every botanist has made a contribution to the debate, but the distinctions postulated by O. Kuntze are now the accepted ones. *Camellia* and *Thea* are regarded as members of the same genus – *Camellia* – belonging to the family Theaceae. Widely different ecotypes exist, notably the bushy *sinensis* and the rangy tea trees of Assam, which also grow wild in China. There are about 240 species in the Theaceae family, of which about seven produce some sort of infusion. Only two of the species, however, the bushy and the rangy types, have any real economic importance.

10 Short ton = 2000 lb; metric tonne = 2200 lb; long ton = 2240 lb.

11 Ironically, the world's best supply of kaolin, which is now used in many other process industries as well as china, is found in Cornwall, and the residual dumps around St Austell give the landscape its peculiar character. Cornish kaolin (china clay) is today exported all over the world, even to China, but these deposits were unknown before 1750.

12 Alice Hanson Jones' seminal *Wealth of a Nation to Be* provides an excellent picture of the true wealth of the colonists in the 1760s and 1770s.

13 The stamp tax was levied on all legal transactions involving real estate, partnerships, shares etc., usually at about 1 per cent of the sale/purchase or succession or whatever. The tax was difficult to escape, and therefore doubly unpopular.

14 The Navigation Acts, only repealed in the late 1840s, decreed that trade into and out of the United Kingdom should only be conducted in British ships, or vessels of the country of origin of the imports (or country of destination in the case of exports). This restrictive practice was intended to strengthen the Merchant Marine.

15 The Indian laws, of various dates, prohibited the sale of intoxicating liquor, firearms or explosives to the Amerindians. As every connoisseur of westerns is aware, these laws were honoured more in the breach than in the observance.

16 From 1660 to 1914, with a short period of suspension during the Napoleonic Wars, England (and later the United Kingdom) was on a standard of currency value which related the pound (sovereign) to a known amount of gold. This was the 'gold standard'. It gave the British economy a stability and discipline which proved too irksome for more recent politicians. Since the Bank of England stopped gold payments in 1914, the price level in Great Britain has inflated more than 100 times. Between 1660 and 1914 the price of wheat, for example, remained static within a range which reflected the state of trade, harvests and surpluses, and which was free of long-term inflation.

17   The importance of tea, even a hundred years after the Boston tea party, can be appreciated from the widely known and practised habit of coffee adulteration. Between 1859 and 1875 genuine coffee doubled in price in real terms in New York. *The American Grocer* of 29 April 1876 gives the following as an excellent basis for 'ground coffee' retailing at 25c per pound, at a gross profit of 400 per cent:

Roasted peas	40%
Roasted rye	20%
Chicory	10%
'Other'	5%
Best coffee	25%

'Other' included tallow, depending on the greasiness of the chicory, which was always to be pre-mixed with the dry rye before grinding the whole concoction. All this adulteration took place before the massive imports of Brazilian coffee made the task too much trouble for too little gain. But even in 1940 a great deal of 'coffee' being served to the public was essentially adulterated. A hundred years previously, 'real' coffee had been a great rarity. Most coffee served outside the homes of the rich was roasted corn. Slaves, servants and children, served real coffee as a treat on Christmas Day, were often very unwell, suffering from the effects of unfamiliar caffeine.

18   pH is a scale of acidity/alkalinity with the number 7 representing neutrality; 5.0–5.5 is moderately acid – suitable for azaleas, rhododendrons and even potatoes, but a poor soil for good cereal crops.

19   In the original Linnean classification green tea and black tea were thought to be two separate species (see note 9).

20   Cut-tear-curl machines were first produced in the early 1930s. The original machines consisted of two engraved metal rollers, rotating in opposite directions, like a mangle, but feeding the product through in a continuous flow. One roller would rotate ten times faster than the other – typically at 750 and 75 rpm. The coarse leaf which has been withered and rolled once is put through the CTC machine, a process which breaks the leaf's cells without loss of juice or rise in temperature from friction. The end product makes a strong infusion, brisk but harsh. The method greatly improves teas picked in the rain. Later, other machines were designed to do the same job. All the machines produce a reliable, second-class tea, which may be more profitable than producing very variable teas; in other words, CTC machines produce reliable mediocrity.

21   The extraordinary and highly successful city of Hong Kong, itself a child of tea, is an indicator of how urban China might have turned out had it not been for the Anglo-American use of opium as a payment mechanism, and China's consequent destabilization.

22   China had no copper deposits of its own; for the wealthy, silk was the only material that was comfortable to wear in the hot, humid Japanese summer.

23   The whole question of guns, gunpowder, Europeans and Christianity is discussed at length in Noel Perrin's *Giving Up the Gun*.

24   Perry arrived in Japan in 1854, and signed a convention which apparently opened Yokohama to American traders. He had, however, impressed the reluctant Japanese with the overwhelming nature of his naval force rather than the virtue of his economic arguments. More than ten years elapsed before there was any appreciable trade between the USA, who were keen, and the Japanese, who were the reverse. It was not until 1866 that the tariff duties fixed on imports were reduced from 15 to 5 per cent. At no point since has Japanese foreign trade been permitted to diverge from what has been conceived to be the national interest.

25 In terms of firepower, vehicles, aircraft, even food consumed, the Americans used nearly twenty times as much crude energy in defeating Japan as Japan used in defending herself. In the end, that twenty times would have been massively exceeded but for the successful explosion of the two atomic bombs which arguably saved another year of conventional warfare. Those two bombs, however, cost more energy than was expended in warfare in Egypt and Libya in the whole of 1940–2. Profligate Westerners do not yet understand the importance of Japanese energy efficiency in the whole matrix of Japanese industrial success.

Cotton

Cotton
and the Southern States of America

In 1784, a year after the Treaty of Versailles had legitimized the independent United States, the first American cotton arrived in the port of Liverpool. It comprised a single bale. Under the Navigation Acts, raw materials
were only permitted entry into Britain in British ships or in vessels belonging to the nation which had exported those goods. The Liverpool customs
officers refused to believe that the bale of cotton could have come from the
USA; they were convinced it must be from the West Indies, and was therefore of British, French or Spanish origin. Since it had been brought in an
American ship they refused it entry, and it remained on the quayside until
it rotted.

One of the people taken to see this bale of cotton was a nine-year-old
boy, the son of a merchant. The child grew up to become a noted cotton
importer and lived until the autumn of 1861, when he was eighty-six. In
his last spring he exerted himself writing to his agent and associate in New
Orleans to point out the folly of the Secession of the Southern States from
the Union, which was to result in the Civil War, and to urge compromise.[1] His attitude was unusual, since many of the so-called 'Liverpool
Party' were in favour of an independent South and, probably, the continu-
ation of slavery.[2]

That man's life encapsulates this chapter, the story of the explosion of
the American cotton trade, which grew in his lifetime from one bale to
4 million, and includes the whole history of Southern slavery and the
American Civil War.

America is the country it is because of its past, like any other country.
But because nearly the whole of American history is documented, we
know far more about the origins of the white American nation than about
any other world power in history, more than about the roots of any European country, more than about contemporary Japan or Russia.

The years 1784–1861 were critical for the American South, and for the

North of England; they encompass a period from the year of Dr Johnson's death and the drawing-up of the American Constitution, to the world of 'Victorian' endeavour. On the way, cotton played a major part, from a pre-steam age in England and an American Jeffersonian idyll to dark satanic mills and black slavery, from sail and waterpower to steam and iron, from cottage industry to industrial technology, from the potential of the free black minority to an enormously increased and recessive slave population, from two countries, mother and daughter, to the same two after each had undergone a different revolution.

Before the American War of Independence, cotton was a labour-intensive crop which cost many more man-hours than the alternatives.[3] By 1861, the cost of industrialized cotton cloth in Europe or the USA in terms of gold had fallen to less than 1 per cent of its cost in 1784. This was a revolution in true cost, the speed of which the world had never seen before: the only comparable modern equivalent has been the reduction in the cost of nylon goods since 1945. The cost to producers and converters of cotton was of course very high, and in hindsight would anyone willingly have paid the price?

Man uses all sorts of vegetable fibres – flax, hemp, jute, manila and sisal are the common ones that have always been important. More unusually, the pineapple plant yields a very fine thread after an infinity of labour, and a coarse cloth may be made from nettles. The mulberry leaf, after pre-processing by the silkworm, can be converted to a fibre of unique luxury.

Cotton fibre is flat and hollow, but already twisted; under the micro-scope it looks like a canvas hose, empty, flattened, but with a natural twist. The staple[4] length varies from that of natural, wild cotton, with hairs only ⅓ inch long, to the selected or hybridized cottons of commerce which may reach nearly 3 inches. By and large, the longer the staple, the easier it is to work, and long-staple cottons produce 'silkier' thread, which is a more valuable product.

Within the landmass of Europe, Asia and Africa, cotton appears to have originated in Egypt. It passed westwards and eastwards, reaching Spain in about AD 900 and China, Japan and Korea in about AD 1300. By the Renais-sance supplies were available in the countries with a Mediterranean sea-board, and Europeans took cotton to cultivate in the Eastern Atlantic islands – the Azores, Madeira and the Canaries. At that time, cotton does not appear to have been grown in non-Mediterranean Africa. Cloth was imported from India, from Persia and even from the southern Arabian peninsula.

On arriving in the New World, the early explorers found several diffe-rent species of the plant, and in Mexico a handicraft industry of considerable size. In 1519, Cortes had been presented by the inhabitants of Yucatan with a highly decorated, gold-encrusted cotton robe; this would seem to indicate that the indigenous Mexicans had a cotton textile industry, but also that the cloth was considered a luxury, as with all other ethnic (pre-industrial) cotton cloth at all times.

Thousands of words have been written by botanists and historians about the origins of the cotton culture of mainland America. There is little consensus, except that no wild variety is native to what is now the United States, and that cotton – *Gossypium* – covers a huge range of plants. There are said to be ten species altogether, and within these ten many local varieties and hybrids, since *Gossypium* hybridizes with ease. Two main species are native to the Western Hemisphere, but many varieties exist within those two. Each species is divided into perennials and annuals, into frost-tolerant and frost-sensitive plants, into green- and black-seeded plants, into long-day and short-day plants,[5] into salt-loving and salt-hating, drought-resistant and high-watertable lovers, and so forth. Some wild cottons grow in South America at altitudes as high as 5000 feet; others like saltmarshes. The West Indies were probably colonized by cotton plants from mainland South America, perhaps traded by the early Amerindian settlers in Barbados, and then brought by some person or persons unknown – though it is likely that they were European – from the West Indies to Virginia, Georgia and the Carolinas. Within the mainland colonies, certain varieties settled down to become long-staple, silky, perennial, frost-hating, salt-loving 'Sea Island' cotton, while other varieties became the upland cottons, usually annuals, short-staple, dependent upon generous spring rainfall, and so on.

Cotton requires a deep, well-drained soil rich in humus. It also needs natural or artificially provided moisture equivalent to about 4 inches of rainfall per month during the critical first three months of growth, and then much less during the long picking season, which lasts more than 100 days. Absence of wind is important, since the bushes are unable to survive gales. If frosts have ceased in April, upland cotton is planted in clean, well-cultivated ground, kept weed-free by hoeing; picking starts when the first bolls (flowers or fruits) set in July–August, and continues until the first frosts, since the 'flowering' and 'fruiting' constitute a continuous process. Drought restricts growth, and while a sudden drought in June will hasten the harvest, only a small crop will be produced. Plants of most varieties would survive more than one year, but growers tend to replant every year to prevent a build-up of pests and diseases.

Before the invention of the appropriate machinery, cotton was a very labour-intensive crop, not only in the growing but also in the harvesting and, above all, in the ginning. Ginning is the separation of the cotton seed from the hairs, which form the lint that goes to make the textile; there is three times as much weight of seed as lint. The seed, a very valuable source of crude protein for animal feed and oil for the food industry, was totally neglected, except as a fuel source, until after the Civil War. The simplest gin is a board into which nails have been driven. When the raw cotton is dragged across the gin the seeds stick to the nails, while the fibres are freed to be collected and baled. Ginning by hand is a laborious task – one motivated man might gin a couple of pounds a day; the norm for a slave was less than a pound.

After ginning, cotton must be cleaned and carded, carding being the process by which the fibres are made parallel. Cleaning – the removal of dirt, detritus, bits of leaves, small pieces of twig, and short and unusable fibre – reduces the weight of the ginned cotton by 50–75 per cent. The fibre is now ready for spinning and weaving.

Before 1800 cotton reaching Europe came not from the United States but from Brazil, the West Indies, the Middle East and India, as well as from pre-Renaissance suppliers in Egypt and the Mediterranean littoral.

The extraordinary amount of labour involved in the process from farm to consumer is what made ethnic cotton so expensive. It was comparatively easy to grow a good crop of cotton, but picking 100 lb of bolls would take 2 man-days; ginning would take 50 man-days at best; and baling by hand, cleaning and carding another 20 man-days. All this effort resulted in only about 8 lb of spinnable cotton, which would then require 25–40 man-days to spin.

An ethnic cotton thread, therefore, cost 12–14 man-days of labour per pound. Even if some of this was cheaper child labour, or adolescent labour, the cost of this boring work was still high and inescapable. Compared to 1lb of cotton, 1lb of wool at the same date took at most 1–2 man-days from raw material to thread, linen 2–5 and silk about 6. No wonder cotton was the luxury cloth in 1784.

Labour costs were higher in Europe than in the Middle East or India, and there were high tariffs on imported cloth, so a halfway house between the raw material and the woven finished product was preferred: cotton thread was the favoured pre-industrial import from the East. Bales of 'cotton wool' were also imported – a risky undertaking, since by its very nature a bale of cotton is apt to allow the producer to obscure its true contents. At that time ethnic trade in every commodity was very subject to adulteration and there was negligible quality control. Iron filings were added to tea, salt to coffee and sand to flour – the price usually reflected the risks. So a bale of cotton might contain earth, seeds, cheaper fibres, cotton leaves, twigs and unpicked bolls. On the whole, thread was a safer buy.

★

For some reason not wholly clear, while women did the spinning of ethnic cotton, weaving was a man's job. The problem in pre-industrial weaving was the quality of the warp, which consists of the lengthwise threads on a loom, against which the weft is woven.[6] Until high-quality spinning machinery arrived, no cotton warp was strong enough, so warps were traditionally made of wool or linen in Europe, and of silk in China, India and the Middle East. A cotton cloth of any durability therefore had to wait for this technical innovation; fine cotton textiles such as muslin, shawl material (Paisleys) or fine kerchiefs were nearly always imported in the eighteenth century. The extraordinary added value of the finest muslin made in Damascus was quoted in 1770: 'A fine muslin, made from a lb of cotton wool costing 4d, would be worth £2 as fine yarn, and £10 as cloth, and £15, if ornamented by the children in the Tambour, the process yielding a return of 900 times the cost of the lint.'[7]

The prospect of achieving such extraordinarily high added value, with some large part of the margin being profit, was what drove men to work out methods and means to bring down the true cost of cotton cloth. The textile revolution in the eighteenth century was primarily a cotton revolution, since the mark-up in ethnic woollen, linen and silk manufacture was never as high as the value added to cotton in conversion to ethnic cloth. On the other hand, the price of the raw cotton was always much cheaper than the cost of raw wool, silk or linen. So the conversion process was rightly deduced to be the problem, and the argument was always directed at cheapening the process between raw cotton and cloth.

The Industrial Revolution is usually perceived as the addition of water- or steampower, canals, roads and railways to what had gone before. In fact it was more a revolution of management and method, and that particular charge occurred before mechanization. Men of money and brains recognized the inefficiency of the whole ethnic process, with perhaps all the members of a single family performing every job in the conversion of cotton wool to cloth. Specialization had to come before machinery; ingenious hand improvements were allied with the division of labour; there had to be enough capital to buy the workmen the right tools: these all arrived long before any satanic mills.

The organization of pre-industrial cotton manufacture had started by about 1720 in England and pre-dates by at least two generations the adoption of any of the great power-driven inventions. At least 150,000 men, women and children were employed in Lancashire, Cheshire, Derbyshire, Nottingham and Leicester in 1750, before the first watermill or steam engine was built. But these people would usually work part-time with their family and a few neighbours, employed for some 100–150 days a year in carding, spinning, weaving and finishing the cloth at home. Small, specialist, purpose-built workshops date from the 1720s, and gradually outwork at home gave way to inwork at the factory (or manufactory as it was called); this process of the division of labour, much lauded by Adam Smith, was often fully developed locally before power was applied to the processes involved in cotton manufacture.[8]

Two revolutions in technology came about before any steam- or water-power was added to the cotton industry. The first, in the 1730s and 1740s, was conducted by dilettante investors and talented but inexperienced 'mechanicks'; it was unsuccessful and is largely forgotten. The second revolution took place in the 1770s; it was pursued with ruthless concern for profit, and led to a growth in the demand for raw cotton which did not level off until after the American Civil War. Over a twenty-year period between 1770 and 1790 most of the features of the modern cotton industry were established.[9]

The earlier developments of the 1730s and 1740s are intriguing because of the close involvement of Dr Johnson and his friends. The key figure is one Lewis Paul, probably the son of a Huguenot refugee who had settled in Spitalfields, the silk-weaving area of London. Paul was a wild young man who wasted his youth in riotous living, after which he became an apprentice to a maker of grave-clothes.[10] Always seeking amusement and conversation rather than work, at an early stage he exhibited the characteristic of the intelligent in an uncongenial occupation: lighten the workload to lessen the tedium. His gloomy trade involved finishing off the clothes with pinking all round the periphery of the shrouds. Paul first invented a pinking machine, and then very rapidly became intrigued by the problem of spinning cotton thread.

The high labour cost of producing cotton cloth anywhere was caused not only by the problem of removing the seeds from the lint, but also by that of spinning by hand. These activities could only be carried out in England at all because of the high rate of duty on foreign cloth and yarn, a form of protection introduced in 1722 to combat the high-quality, low-cost product of the Indian village industry, as well as the even higher-quality goods from the Middle East. A weaver would require one man to keep him supplied with warp, another man, or woman, to 'rove' the spun thread, another to card the thread, and five to ten spinners.[11] Spinners were nearly always women ('spinsters'), each using a single wheel with one spindle, and often working in her own home. Each spinster's product would vary, and this variation in quality was as much a disincentive to expansion as was the high cost of ethnic spinning. In the early eighteenth century the spinning process had changed little since prehistoric times. Hand-wheels are still in use in southeast Asia, and can be found on sites dated at least as early as 1000 BC; they are similar to European wheels of the eighteenth century.

Lewis Paul teamed up with a wheelwright, John Wyatt, whose family came from Lichfield and were friends of Dr Johnson – at that date, 1735, newly married, living near Lichfield, and tutoring unsuccessfully only three pupils in three years (one of them was David Garrick). Johnson met Lewis Paul and was enthused by the possibilities of his roller-spinning machine. There is no evidence that Johnson himself invested in Paul's invention, but he had friends and they became investors.

The Paul–Wyatt partnership exploited a patent granted in 1737 in the following way. Paul lived an insecure life as a *boulevardier*, a man of fashion and frequenter of salons, a role much crippled by debt and long periods in a debtors' prison. John Wyatt, on the other hand, stayed in Birmingham and built the machines. Paul would persuade Dr Johnson's friends to lend him money, ostensibly to finance machinery; in the end none of it was repaid, but in lieu of money Paul granted 'licences' for the operation of so many spindles. Wyatt frequently sued Paul for the money for the machines, and the investors also sued Paul, so he was sent to prison for debt every two years or so. Everyone complained to Johnson, who began to regret his sponsorship. Most of the 'investment' appears to have gone to support Paul's extravagances.

What might have become the foundation of a great industry died with Lewis Paul in 1759. Meanwhile, his spindles had been installed in Northampton, Birmingham, London, Leominster and Lancashire. Altogether perhaps fifty workshops were built, with a total of 2500 spindles, but the owners of these enterprises were not textile magnates: they were Johnson's friends, authors, booksellers, printers and publishers, and this literary invasion of the textile trade failed, as might have been expected. But though the development was buried with the entrepreneur, nothing can take away from Lewis Paul and John Wyatt the historical credit for having devised a successful spinning frame.

The second successful revolution is associated with the brutal but effective figure of Richard Arkwright, a man with a notable talent for business and no great inhibitions about nobbling the competition. He was the archetypical early self-made man, an organizer, manager and manipulator of genius; he was also one of the first generation of industrialists, as opposed to bankers or merchants, to be knighted. Arkwright began as a barber, and built up his first modest fortune by conserving the human hair from his own and other people's barber's shops, treating it, dyeing it, spinning it into longer lengths, and selling the product to wigmakers. Many men have entered the textile trade from odd places, but few have started out by reconstituting the sweepings off barbers' floors.

When he was thirty-five Arkwright, now modestly wealthy, turned his attention to the problems of cotton-spinning, and in 1769 patented his spinning frame which, unlike other frames, would spin not only the weft but also the warp. This represented a major technological breakthrough in cotton-spinning. For the first time a pure cotton cloth of great durability could be produced, but it would be twenty years before it was commonly manufactured. Arkwright's frame needed improvement and exploitation, and he was uniquely equipped to do so.

In 1770 he built his first mill in Nottingham, powered by horses rather than water. It contained 2500 spindles, as many as Lewis Paul and John Wyatt had built in a generation, and did 2500 homebound spinsters' work with only 50 unskilled operators. The next year he built his second factory, larger and powered by water, at Cromford in Derbyshire. Between 1771 and his death twenty years later Arkwright, acting with various partners, erected mills (all water-driven) at an average of one a year. But there were problems to counter these successes. He was much engaged in court actions brought against him for his alleged habit of infringing patents; he lost most of these cases, the prosecution claiming that he had copied the inventions of others, that he was deliberately obscure, or that he had profited so much by his enterprise that a patent was unnecessary. The first mill to be destroyed by irate displaced skilled labour was one of Arkwright's, at Chorley in Lancashire in 1779. He was the victim of the first transatlantic piracy, when some models and parts of some of his machines were smuggled to New England in 1786. He also suffered all his life from chronic asthma of nervous origin; when thwarted, he could hardly breathe. The heart attack that killed him in 1792 was induced by a shortage of breath aggravated by asthma, initiated in turn by a dispute with a customer.

The weight of cotton spun in England rose from less than 500,000 lb, all spun by hand, in 1765 to 2 million in 1775, mostly spun by machine, and to 16 million in 1784, all spun by machine. Before about 1790 the mills were all water-driven, and they were well spread. In 1788 the distribution was: Lancashire 41, Derbyshire 22, Nottinghamshire 17, Yorkshire 11, Cheshire 8, Staffordshire 7, Westmorland 5 and Berkshire 2. There was at least one cotton mill in every English county at this date except for Cornwall, Essex, Surrey, Kent, Oxfordshire and Rutland. Water mills were also widely used for silk and woollen manufacture. Waterpower, of course, dictated not only the wide spread of spinning-mills, but also the limitations of the industry's growth. In 1799, nearly 90 per cent of the available waterpower on the upper Irwell in Lancashire was already absorbed.

Traditionally, every English schoolchild has been taught for more than a hundred years that cotton settled in Lancashire and wool in Yorkshire for climatic reasons, the relative humidity in Lancashire being 20–25 per cent higher than that in Yorkshire, and cotton thread more easily handled in a damp climate. But this is less than the whole truth. In 1788, as described above, cotton mills were spread all over the country. At the same date there were four times as many woollen mills in Bradford, Wiltshire as in Bradford, Yorkshire. It was not only waterpower that made Lancashire attractive, but the proximity of the port of Liverpool for the raw cotton imports, relatively good internal communications, cheap coal, cheap iron, large reservoirs of workers off farms and from Ireland, and the local mobilization of capital by people and institutions in Manchester and Liverpool. All this led to the logic of Lancashire, and no other area of England was as suitable for the development of the cotton industry, dampness or no dampness. In any case, nearly all watermills were inevitably damp because of the river whose power they used.[12]

Within two generations a cotton worker had become required to work shifts, often at night, and he was frequently obliged to live in a company-owned 'back-to-back' without sanitation, garden or fresh air. His situation was worse than that of his grandfather and arguably worse than that of the slave who grew the cotton in the Southern States for the factory in Lancashire.[13]

Enclosures, whose final chapter dates from the beginning of the Industrial Revolution, have often been blamed by Radicals from William Cobbett onwards for the misery of the working classes in industrial England,[14] but a more worthy target might be the concentration of labour in factories. The opportunity had now ceased to live in a cottage halfway up a Derbyshire valley, engaged, with a wide circle of friends and relatives, in following a complete series of operations from cotton wool to garment: spinning, carding, roving, warping, weaving and making up, piece by piece, day by day, at the unhurried pace of an individual.

By 1861 the management of the specialized division of labour had separated the various operations into mills perhaps 20 miles apart, linked by railways, and driven by steam, but requiring the operatives to live within walking distance of the mill. The productivity of each element of the process had risen by between ten and fifty times, for a relatively low cost.[15] The whole industry had expanded to absorb not only all the displaced piece-workers (except in bad times), but also many thousands more whose grandparents and great-grandparents had had nothing to do with textiles.

We should not weep for the 'good old days' of cotton cloth. It may have been twenty or a hundred times more expensive in 1784 compared with 1850, but it was also of far lower quality. In general, for daily use, machine-made cotton was infinitely better than the ethnic product.

★

If the period 1784–1861 saw an eightfold increase in the numbers of black slaves in America, it also saw an increase in the misery of the British cotton industry. The 'free' piece-worker, of whom there were perhaps 250,000 in 1784, disappeared; in the process Lancashire became unskilled at work and brutalized in living conditions. In 1825, before any of the really important Factory Acts had been made law, 90 per cent of the workers in the spinning section of the industry were women and children; the children had no real opportunity for education, no protection against abuse, no redress against brutality, no rights in common law against dangerous machinery, inhuman overseers or over-long hours with no overtime pay. In 1784, it could be argued that there was a certain arcadian charm in living and working in a rural hovel with an idyllic setting, however insanitary it really was. The brutal truth of the purpose-built back-to-back house, one of hundreds of identical rows in any manufacturing town, did not lend itself to the idea of charm, arcadian or otherwise. For example, in 1784 the village of Bacup contained about forty cottages whose inhabitants did outwork. In 1861 Bacup had six cotton mills and more than 3000 mill-hands of all ages. The amount of human misery in Bacup in 1861 must have been far greater than in 1784. If the cost to the Lancashire mill-hands was not as high as to the field-hands who lived and died in slavery, the difference must have been one of degree and not of kind. In 1861 Bacup had seven nonconformist chapels to provide the consolations of religion for those who had faith not in this world, but in the next. They were needed, appreciated, and full every Sunday.

In the parallel New England operation in the nineteenth century, only about a quarter the size of the British cotton industry, working conditions were never as bad as in Old England, because an operative in the United States indirectly benefited from the frontier. Though he might not be able to afford to join a wagon train himself, and it required considerable resources to farm anywhere in the USA, the tenant farmer was not forced into industry as in England. The new country had a permanent shortage of labour except during periods of slump, and before the 1840s there were few really poor immigrants. Labour was therefore always better treated in New England than in Old, at least until the wave of paupers arrived from Ireland in the 1840s and 1850s.

Eli Whitney (1765–1825) is one of the folk heroes of the American people. A poor farm boy, he was a self-taught mechanical genius. From the age of ten he would repair or renew anything made of wood, iron or leather: spades, wheelbarrows, harness and the other simple tools and equipment of colonial Massachusetts. In 1788, Whitney crossed the state boundary into Connecticut and became a student at Yale University at New Haven.

At that date there were only about a hundred undergraduates and two professorships, one in theology, and the other in 'Mathematicks and Natural Philosophy', the latter a phrase which in the late eighteenth century encompassed a vast body of knowledge covering such subjects as botany, zoology, chemistry, physics, agriculture and geology. Yale also insisted on the study of the Bible, and on the Puritan faith; the affairs of this world and the next were well learnt. Whitney met the cost of his fees by exploiting his skill as a mechanic; on one occasion he spared the college authorities the trouble of sending an astrolabe to London for repair, also saving them £15 – about a year's fees for Yale at the time. He was intrigued by problems of a mechanical or mathematical nature, almost as good with his brain as he was with his hands, but intensely shy, reserved and not at ease with people.

After graduation, he was persuaded by a Southern friend to go to Savannah, Georgia to take up a teaching post. The job never materialized, but the lonely young man was befriended by the widow of a Revolutionary general, Mrs Nathaniel Greene. In Savannah for the first time he became aware of the problems of cotton, and in particular the problem of ginning, which was done either with a crude, nail-studded board or, worse, by hand. It was typical of the times that none of the fine gentlemen at Mrs Greene's table, or the white overseers in the field, or the many people involved in cotton in the port of Savannah had ever set their hand to cotton-ginning in their lives. The whites in the Deep South left work to slaves.

But Yankees from small farms in Eli Whitney's Massachusetts had to use their own ingenuity to save themselves time and enable them to perform the same task every day with maximum possible efficiency. There is no evidence that Whitney ever considered working in the heat and humidity of a riverside plantation, but the idea of a simple device to increase the productivity of the slaves, to remove the bottleneck of ginning and to accelerate the growth and marketing of cotton teased his mind for nearly a month before he came up with the answer.

There is a famous legend that Whitney was permitting his subconscious to mull over the ginning problem as he idly whittled a hickory stick in the back yard of Mrs Greene's house, when suddenly he saw a cat strike out at a chicken and catch in its claws not the chicken, but only a few feathers. The idea of the mechanical gin was born. Whether or not this fable is true, it is a good story, and the Whitney gin was built in three days. It is more likely that Mrs Greene and Eli Whitney had worked on the problem after the young man had built various household devices such as draining boards, settling pans and an early steamer. A graduate in natural philosophy would be likely to use a more methodological approach than the flash of inspiration beloved of poets – the cat and the chicken belong in the same class as Isaac Newton's apple.

The Whitney gin consisted of a solid wooden cylinder into which headless nails, set half an inch apart, had been driven in an ordered pattern. On the outside of the cylinder was a grid, with bars set so closely together that the seeds could not pass, but the cotton lint was pulled through by the spikes. A revolving brush cleaned the spikes, so that at each revolution the seeds were detached and fell into another compartment. The machine was worked by hand, and with it a slave could gin not 1 lb but 50 lb of cotton a day. To a rational man in a rationalist age, it must have seemed that he had benefited mankind and that his fortune was made.[16]

But Whitney was to learn the hard way that an inventor is the prey of the market. The first model gin was stolen. A second was made but before the patent was granted on it, in March 1794, many copies of the first model were in use. This simple mechanical device, which any wheelwright, blacksmith or carpenter could make, spread like wildfire throughout the South. Whitney formed a partnership with a man called Phineas Miller, who was courting the widow Greene and subsequently married her. Miller and Whitney built a factory in New Haven, Connecticut,[17] near the university, to build the gins and sell them not for a fixed price, but for a third of the value of the cotton which went through the gin. The factory burnt to the ground; it was rebuilt. The demand for gins exceeded the supply. In May 1796 one Hogden Holmes patented a copy of the Whitney gin, using circular saws instead of spikes. Whitney spent the time he ought to have employed in the manufacture and sale of gins in prosecuting infringements of his patent, and in 1807 the validity of the original and all subsequent patents was finally settled. But by then, everyone in the South could obtain a gin of sorts locally. Perhaps such a simple device could never be protected by patent in those relatively primitive days. Certainly, no contemporary patent on spinning or weaving in England successfully withstood infringement; one might as soon have patented the waterwheel.

The cotton South, which benefited so much from the gin, made some kind of restitution to Whitney. South Carolina voted $50,000 from state funds for the infringement of the patents; North Carolina imposed a tax on cotton for five years, which yielded $30,000 or thereabouts; Tennessee gave him perhaps $10,000. Only Georgia, a frontier state with frontier business ethics, but the state in which Mrs Greene lived, where the cat missed the chicken and where the gin was born, failed to pay up. Eli Whitney was naturally disgusted by the whole gin saga, which had absorbed twenty years of his life, yielded him less than $100,000, and made other men rich beyond the dreams of avarice. He was to go down in history as the essential link in the whole cotton saga, one of the godfathers of the War between the States.

By the beginning of the nineteenth century, when Whitney gins, or copies, were freely available, there was no barrier to the growth of the cotton kingdom. The slaves who were displaced by the gins could be put to some task not yet mechanized. It is hard to overstate the importance of the coincidence of the escalating demand for cotton from England combined with the increased production made possible by the new ginning process, whether Whitney's or that of some imitator.[18] The soils of the Old South[19] were as exhausted as their owners and as poverty-stricken. Slavery had gone out of economic fashion in Virginia, Maryland and Delaware. Tobacco was no longer accepted in England without heavy duties. Trade in meat, wheat and corn with the British West Indies was made difficult, and sometimes impossible. Without the opportunities opened up by the gin, the Old South might well have sunk into decline, and the Deep South never have been born a child of slavery. There was, of course, a certain historical inevitability about the invention of the Whitney gin, though if it had been delayed by twenty years the whole history of the United States might have been different. But as it turned out it was cotton which moulded the nineteenth-century South, and the gin was one of the factors which made the cotton kingdom possible.

Cotton acreage in the New South was to rise to match demand, and with each 100 acres of cotton between ten and twenty slaves were needed. The Old South, its land unfit for cotton, became a breeding-ground for slaves. The price of a field-hand, which had fallen in the period 1775–1800 by half, began to go up; in the fifty years between 1800 and 1850 a prime field-hand rose in price from $50 to $800–1000 in real terms. The slave trade, which had been an export trade in the 1790s because of the agricultural depression in the South, became the single most important commerce between states, and illegal importation went on from the West Indies, from Africa, and through Mexico and Texas.[20]

The case of Eli Whitney's gin illustrates a truth that took thousands of years to penetrate the human consciousness and which did not become part of the established wisdom of the cotton states until the Civil War. Labour is unpopular with all human beings, and unnecessary labour, if recognized as such, is the most hated. Yet in an unmechanized slave society no one had much interest in saving labour, since an exhausted slave gave less trouble after sunset. Like a mule or a horse, he had to be worked until all thought of misconduct had been driven away by fatigue. And if a slave did a particular job quickly, he was merely given a second or third task. For ease of management, slaves tended to be worked in gangs at the speed of a leisurely amble, slowly enough to avoid exhaustion, quickly enough to avoid the whip. So all the cheap, clever ingenuities which the self-employed labouring man, the artisan or the piece-worker uses to increase productivity were unknown in a servile society. In this sense, the cotton kingdom was a reminder of what life had been like for the labourer in Ancient Egypt or Rome, or during the Middle Ages.

The self-employed artisan, so much admired by Thomas Jefferson, grew in strength and political importance in the North, not the South; and on the frontier, not in the settled regions. Even in a matter so vital to the South as the cotton gin, it was Yankee ingenuity which solved the problem. When a Southern cotton-grower wanted to increase production, he bought more land or more slaves; these resources cost money. A Yankee usually had no money to spare, and for the advancement of his life and hopes he had to use his own brain. This cultural difference was to affect the course of the Civil War in the 1860s.

If one were asked to characterize antebellum Dixie just before the first shells fell on Fort Sumter in April 1861, one might think of Tidewater plantations growing rice, indigo and sugar, and seaports of sophistication, fashion and lechery.[21] Inland lay the Piedmont section, with gracious mansions and educated ladies surrounded by luxury, flirting with elegant, raffish gentlemen of quick temper and ardent chivalry.[22] In the mountains, poor whites would be scratching a living from gold diggings, timber extraction or subsistence farming, but eager to join the Confederate Army. The whites in the South were overwhelmingly Anglo-Saxon Protestants, the essential supporters of a way of life which – so the North thought – no one could defend.

In fact, of course, these images are no more reliable than the first chapters of *Gone with the Wind*. Cotton certainly made the South apparently rich, and made slavery apparently inevitable, and made the combination apparently unbeatable. The reality was less certain and more sordid.

Cotton did not overtake tobacco in value within the USA or sugar in world trade until some time after 1820, when the volume was 200 million lb or less than 400,000 bales. In 1861, production was nearly 2000 million lb or 4 million bales. Though the price had fluctuated, volume of production made cotton very easily the number one crop, the king. The number of bales produced roughly equated to the number of slaves, including the servants, indoor and outdoor, the blacksmiths, the carpenters, the cooks, the nurses, the retired, the children and the field gangs upon whose sweat the whole edifice rested. Each black 'produced' just over a bale of ginned cotton. But this picture of a mature rural civilization was not always true. Before 1820 most of the cotton was grown east of the mountains, by whites rather than blacks, in ordinary crop rotation.

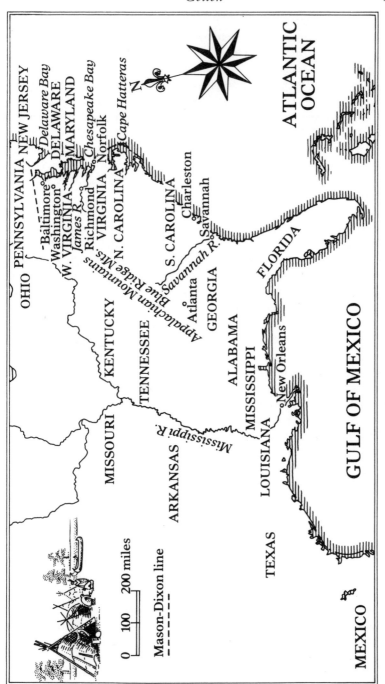

The South-eastern United States

The spread westwards and southwards from the tidal South followed a pattern from the very early days. It was limited by terrain, government and Indians. The Tidewater section had long, navigable inlets which permitted trade into the interior without roads. There was and is a long, navigable, sheltered channel from New Jersey to Cape Hatteras. Particularly in and around Chesapeake Bay, the James River, north Delaware Bay, and south beyond Norfolk, Virginia, traffic was by water, and mostly by sea. This section also traded directly with Europe, often from their own wharves, without much reference to customs officers or any other paraphernalia of bureaucracy. 'A ship might be tied up to a tree in Carolina, and a month later be in the Pool of London,' said John Locke in 1669 when discussing government control.

The government – the British-appointed governor, or the state legislature, or, in due course, the federal administraton – found it difficult to control export and import, and the activities of the rich Tidewater planters who grew tobacco, rice and indigo and wanted free trade with each other and with Europe. The government also found difficulty in controlling those who streamed westwards and southwestwards. The history of America is strewn with examples of rules and regulations drawn up years after the event. In trade with the Indians, many governments banned the sale of firearms and liquor years after these objects had been shown to be what the Indians wanted. The Spanish failed to prevent the Indians acquiring horses. The British failed to prevent the adventurous, the shiftless and the independent settling inland from the original Tidewater settlements. Not only did this ex post facto legislation not work, and not deal with the real world, but the mere effort brought the law into disrepute and gave the frontier an undeserved reputation for lawlessness.

Even before there was any considerable 'West', there was already a 'frontier' and problems of the indigenous natives to be faced. While it is fashionable today to decry the white settlers' treatment of the Indians, treachery, violence and deceit were practised by both sides: by Indians against each other; by Indians in company with the French against the British and the colonial Americans; and by Indians in company with the British against the French. There is not a square mile east of the Mississippi which has not witnessed a scalping or a murder, a massacre, a trick or an ambush. The inherent problem was that the whites wanted the land, and would use it, they thought, more efficiently than the Indians; the Indians, on the other hand, wanted to stay put. If the Indians resisted, some circumstance, concerted or unorganized, would be needed to free the land, or at least to allow settlement. The indigenous peoples would have to be separated from the land they occupied. However viewed, it is difficult to see how the aims and ambitions of the Indians could ever have been reconciled with the restless energy, land hunger and manifest destiny of the American people.

The distinct feature of American life is not the nobility of the Founding Fathers, nor freedom from the arbitrary rule of kings, nor escape from the tyranny of Europe, both secular and religious. None of these desirable attributes would ever have been achieved if the key had not existed: a relatively empty continent. No one knows what the Red Indian population was before the arrival of the whites. But even taking the highest respectable estimate, about 25 million, still makes the United States the emptiest fertile country in the world when the white man arrived.[23]

Cheap, virtually empty, fertile land had more influence upon the early settlers than all the rhetoric of all the politicians who have ever inspired, amused or saddened an American audience, and has shaped the American character more than any other factor. For the first time in modern history, land was available for about one-fiftieth the cost of the same quality of land in Europe. Sometimes it was virtually free, stolen from the Indians at the risk of revenge; sometimes it was bought for a few baubles, or a bottle or two; more often, it was settled with the authority of the federal government, as were nearly 10 million acres by 1820. The very poor could never aspire to become settlers in the West or the New South, but others improved themselves beyond expectation in America. Younger sons found more cheap land than they could in the Old World, as did tenant farmers. In the United States small shopkeepers and other men dispossessed by change were not pauperized, made landless labourers and driven into towns to become the new proletariat, as in England.

The reverse of this freedom was that virgin land became much less valuable than the improvements made to it. The forests were exploited and then burnt to clear the way for cropping; the amount of labour put into the land, slave or free, became the measure of its value; investment in slaves to work the cotton fields was between five and ten times the investment in land, clearance included. The land itself, being very cheap, was not treasured: husbandry suffered. Output per man-hour was always more important than output per acre. Practices inherited from Europe and used with success in the Tidewater for years were neglected for quick returns, high profits in the short term, and the optimum use of labour, scarce if white, expensive if black. The older generation disapproved. 'We are bad farmers,' said Benjamin Franklin, 'because we have so much land.'[24]

From the earliest times, the white man in the South pushed inland. There was always a better opportunity somewhere else, and a belief in the sovereign remedy of a change of scene, a belief which has had a profound and perhaps unsettling effect on the American psyche. Before the Civil War this belief led to a steady migration from the North to the northwest, and from the Old South to the southwest. The direction of movement of the feckless, the ambitious, the troublesome, the energetic, the enterprising, the debtor, the indentured servant free of his master and the petty capitalist was away from the Tidewater region; these people and many others went west. What none of them had before cotton was any kind of cash crop.

The economic history of the pre-cotton American South was one of constant debt. For nearly two hundred years after the first settlement until the early 1800s the South produced nothing which Europe could not find anywhere else. Tobacco was grown all over the Mediterranean and the Middle East within a century of its first being imported into Europe. Indigo was available from any subtropical country. Rice was often cheaper from the Mediterranean or the Far East. Sugar was hardly grown in the USA before the Louisiana Purchase in 1803,[25] and never competed with the Caribbean in the export trade. Grain was so bulky and transport so expensive before railways that its culture was limited to areas close to rivers or the sea, and except in times of European shortage it was more usually exported to the limited market of the West Indies.[26]

Cotton was unique in the history of the USA for two reasons. The southern United States turned out to be best adapted for the growing of cotton and capable of supplying an apparently insatiable and rising demand. In England, the most important trader, net imports rose from 20 million lb in 1784 (none of it from the North American mainland), to just under 1.5 billion lb in 1850 (82 per cent from Dixie), a 150-fold increase in demand. The American export picture, as opposed to the British import figures, is even more startling. The first bale was experimentally exported in 1784; it was followed by less than 10 million lb in 1800, less than 100 million lb in 1830, more than 800 million lb in 1840, and more than 2 billion lb in 1850. This represented a rise of about 7 per cent compound annually, in a product which, though its price varied, could be exchanged for sterling bills on London or for gold. The debtor South could at least pay the interest on its loans, since in 1850 cotton exports amounted to $80 million; in 1860 they were nearly two and a half times as much. This was more than the entire United States export trade in all other commodities and manufactures at that date, and it remained so until the Civil War. Cotton was indeed king. The importance of cotton, and its known and assured value to the South, can be judged from the fact that if a man in the slavocracy could borrow in London, as opposed to borrowing locally, he would pay on average less than half the interest he would pay in the Southern States. A cotton exporter could borrow in what we would now call the offshore market; neighbours who did not produce cotton could only borrow locally.

Ironically, cotton started as the crop of the poor white. It is unlikely that there were any specialist cotton 'plantations' in 1800; no trace of one exists. Yet in Virginia in 1805, a poor white grew an acre of cotton and sold it for $250, more money than he had ever before held in his hand. Two years later, a man carried some seed on his back 800 miles to Tennessee, grew the first cotton in that state, and floated the product down the Mississippi to New Orleans. In 1823 a man planted some cleared land in Georgia and sold it in May with a cotton crop established; he then cleared some land in Alabama that autumn, planted and sold that farm and repeated the process in Mississippi, finally ending up with 1000 acres of his own, title to which had cost him $1250 and two years' hard work. Many of those who ended up with plantations started in this rough but effective way. Others uprooted themselves, their wives, families and slaves and trekked, as did their contemporaries in South Africa, over the mountains to settle up to 5000 acres in the New South – Alabama, Mississippi and Louisiana. Nevertheless, monocultural plantations as opposed to mixed farms were uncommon until the 1820s, and until that time the coastal lands grew more cotton than either the Piedmont region or the New South, west of the Appalachians.

In any attempt to establish an American pattern in the development of cotton, certain agronomic factors must be recognized. Cotton can be grown anywhere in a huge area of the South, as an annual or as a perennial crop, in rotation or as a form of monoculture. Because of this range of possibilities a constant search for fresh, clean land became an enduring feature of this period, because it was easier than rotating crops. As Thomas Jefferson said: 'It is cheaper for Americans to buy new land than to manure the old.'[27]

The United States in 1820 was an economically vigorous nation, with its population, always on the move westwards, increasing by more than a third every ten years; its industries, mines and agriculture were growing faster than the population but in fits and starts, interrupted by bank crises, problems of temporary over-supply, and shortages of labour in key areas. Until the late 1840s it was also the only white-settled country in the world without a large class of white paupers; destitute immigrants were discouraged. The US population was overwhelmingly Anglo-Saxon and 85 per cent Protestant before the Irish famines of those years. The nation not only grew and grew, but it became more geographically sectional; this sectionalism was itself unique in that it owed little to racial or religious difference amongst the whites.

People in smug, cosy, integrated England, with a most moderate Atlantic climate, and with no place more than 100 miles or so from the sea, may well find it hard to understand the importance of sectionalism in the United States. But to a Frenchman, or an Italian, or even a Swiss, the phenomenon exists, albeit under a different name. In the USA, in general, climate is the most important factor, exacerbated until the jet age by the absence of really cheap, swift and efficient inter-state travel; until the electronic revolution by the lack of any national news-disseminating process; and until electricity was harnessed by the impossibility of any common diet.[28]

In the early days, there was already sectionalism between the Tidewater and the Piedmont areas of Virginia. Diet, climate, economic needs and political ambitions had all created great differences before 1660. This sectionalism was repeated as the frontier moved west, yet remained in the 'Old' areas in the east. Having achieved Independence, the wise men of the young United States were as much worried about national unity, in the absence of war, as about any other problem. Nineteenth-century immigrants became men of New York, New England or the South before they became Americans, or else they obstinately remained Irish, Jews or WASPs. Even in the 1920s, with the breakdown of physical barriers caused by railroads and automobiles, H.L. Mencken was to say that though America was called the melting pot of all the races, in reality the only thing that melted was the pot.

A hundred years previously the pot was in far greater danger, and the sectionalism a chronic deterrent to national unity. States' rights were paramount; 'Americanism' had to be over-stated as a counterweight. Differences between Europe and the early United States became more important than similarities.

The sectionalism which was to result in the house being divided against itself was a function of cheap land, a vigorous population and opportunities, real or fancied, to the west and southwest. The Old South and the coastal region were in decay, with much of the land abandoned and reverted to scrub or wilderness.[29]

By 1820 the United States was settled in the north as far as Lake Erie; the frontier ran across Ohio, southern Indiana and Illinois into Missouri, and then eastwards to the middle of Tennessee, somewhere near Knoxville, and then to the River Mississippi. There were Indian 'nations' in Florida, Georgia, Alabama, Mississippi, Indiana, Illinois and Wisconsin. Independent Mexico owned Texas, New Mexico, Arizona and California. The 49th parallel had yet to be established as the boundary between western Canada and the new nation.

By the time the Old South had been rescued by cotton and the consequent demand for slaves, its most enterprising people had already left. They settled as far north as Kentucky, Tennessee and the southern parts of

Ohio, Indiana and Illinois. They took with them the practice of slavery, but few slaves were taken northwest (though slavery survived in Illinois and Indiana until the 1840s). The lands immediately west of the Appalachians in what is now Georgia, Alabama, Mississippi and Louisiana were to be settled with slaves for the purpose of growing one cash crop only, cotton. There were mixed farms or subsistence farms or peasant holdings occupied by poor whites without slaves, but cotton was the cash crop of the New South, and but for cotton and the opportunities to grow it in the fresh soil over the mountains slavery would have disappeared, as it nearly did in 1784.

In that year the drafters of the Constitution, in their wisdom, had recognized the contradiction between white men who had just wrested their freedom from the tyrannical British monarchy and the black men who were chattels. Nearly all the politicians of the federal period owned slaves, some of them a great many: a few had black or mulatto mistresses. In public they deplored slavery; in private they were slave-owners. In practice, what could be done about the social problem presented by the negro?

Enough trouble existed already in many communities supporting free whites who had been transported from England for some minor offence, served their seven years of indentureship, and then remained there as a drain on their neighbours; the slaves, like the indentured whites, had all their basic material needs supplied and could not be expected, once freed, to know how to contribute to the economy; this situation argued against emancipation. Another argument was that support, tacit or otherwise, for slavery was required to obtain nationwide support for the Constitution. As a result, in 1784 the Constitution-makers rejected the idea of abolishing slavery at that date since they could not reach agreement on this or the compensation issue. In the South slavery was held to be the only way to treat the negroes, who were regarded as a race of feckless children. Its brief decline after 1784 was only for economic reasons.

The beginning of cotton monoculture followed the adoption of the gin in the 1790s; to grow cotton, slavery once again became 'necessary':[30] the alliance between slavery and cotton was begun. In 1808, after which date it was theoretically illegal to import slaves from outside the United States, upland cotton was worth nearly 15c a pound, with the demand expanding annually and the prospects bright. The value, and therefore the price, of slaves rose dramatically. The poverty-stricken but cooler, temperate Old South, consciously or unconsciously, became a breeding-ground to replace the slave trade from Africa and the West Indies; this made the Southern States the first self-supporting, self-breeding slave system the world had ever known. In all previous slave societies the slave population had never maintained itself, usually producing a deficit which was made good by imports, as in the Ancient World or in the West Indies. But the Old South was particularly suitable for breeding and there was plenty of money in the slave-breeding industry.[31]

Human chattelage would now become the most difficult problem facing the whole nation, what the ageing Jefferson called 'A firebell in the night'. True, in the 1790s the French in Louisiana had started a sugar industry which was dependent upon industrial slavery in the cane-brakes. However, that industry was never one-tenth as important as cotton, in terms of either the value of the crop or numbers of slaves; and if slavery had ceased to exist in the USA before the Louisiana Purchase of 1803, it is very unlikely that it would have been permitted to continue in the new lands bought from France.

In the whole of the Deep South throughout the first half of the nineteenth century the only sophisticated metropolitan area was New Orleans. According to European travellers it was also the only place in the whole of the United States with first-class hotels or restaurants. It was a great service centre, an entrepot, a marketplace, a city of leisure, Creole cuisine, theatre, music, lust and vice. Transients were said to exceed the indigenous population in number, prominence and criminality. The number of free blacks, half-castes and those with negro blood 'passing' as whites was said to represent the highest proportion of negroes in the Western Hemisphere; some free blacks were even significant slave-owners.

It was in New Orleans that slave-ownership for reasons of conspicuous consumption reached its highest pitch: negroes to serve at table were matched like carriage horses; pale-skinned mulatto women reached ten times the value of a field-hand, a butler five times, a good cook three times, and the universal little girl who took off the gentlemen's boots twice the value of her less talented contemporaries. The aura of sex and miscegenation, slave music and slave women used for entertainment, the paranoid gambling, universal sharp practice and languid Southern morality which is the folk memory of New Orleans should not obscure the hard economic importance of cotton, upon which all this scum and froth floated. It is rare to find anywhere in the world a city of pleasure without an economically vital hinterland which supports and sustains the fun and games.

More than two generations before steam railroads girded the North, steamboats had transformed the South, and no place more than New Orleans. Before steam, a raft floated down the Mississippi was broken up when its cargo was discharged at New Orleans. It was commercially impossible to sail any craft of any size against the flow of the Mississippi or of its tributaries. The first steamboat made the round trip downstream from Pittsburgh and back up against the current in 1812; the achievement was to puff upstream. Furs, wheat, tobacco, timber, corn, salt beef and many other goods were shipped down from the Mississippi–Missouri basin, and even from upstate New York, Pennsylvania and Ohio. Before railroads, water was the only serious route.

Of all the crops exported from what was then the southwest, cotton was the most important. From St Louis in the north, from Pensacola in the east, from Brownsville in the west, cotton poured into New Orleans for trans-shipment, export, resale, speculation or warehousing. Nearly half the exports to Europe came from New Orleans, easily outpacing the older ports of Charleston and Savannah and the new port of Mobile in Alabama. On each bale of cotton someone somewhere in New Orleans made a buck or two and helped to support the city, with all its garish attractions and dubious charm.

Compared to New Orleans, most of the inland towns of the new Deep South were sordid villages. Apologists would say that men were in too much of a hurry on or near the frontier, wherever it might be, to have time for good manners, but both Yankee American and European visitors made scathing comments about the behaviour of the inhabitants of the South. Compared to the fiction of some novels, the truth is unrecognizable. The white men were generally unclean, the women of any age worn by circumstance and childbearing. The untidiness, the ragged clothes, the assault upon the senses of unwashed poor whites and slaves alike made any antebellum assembly an experience which would disgust the people of an age accustomed to soap, detergents, running water and deodorants.

Even in the 1930s the region provided an adventure for the Northerner, but at any time before the Civil War the Deep South, with the exception of New Orleans, was a cultural shock for the visitor. Malarial mosquitoes were prevalent; yellow fever and typhoid were visiting familiars. Salmonella, worms and dysentery were common, particularly amongst the slaves. For all, black and white, enteric disorder was a constant companion. Heat unrelieved by the benefits of electricity was endured in the summer, along with storms and rain. Ice was only available in New Orleans and then usually only until mid-April, unless shipped at vast expense from the ice-houses of the North. Pure water was at a premium in every built-up area. Sanitation was sketchy and added to the stink and discomfort in an age which was innocent of any knowledge of germs, painkillers or infection. Disinfectants were unknown. Manufactured soap was in short supply. Food, cooked by slaves sometimes disaffected and not always skilled, was variable, often disgusting. Deficiency diseases were endemic in a land of plenty. If this was civilization, it was very raw. On the other hand, survivors of this society were toughened by their experience.[32]

The characteristic problems of slavery were experienced. There was a great deal of petty pilfering, but an ironical lack of sanctions. House servants and hotel slaves, who were the chief offenders in this area, were difficult to chastise or punish; their value would be diminished if they were flogged too hard. They could be sold, but this was a last resort. Cunning slaves judged to a nicety the limits of idleness, petty crime and other delinquencies with which they could tease their owners. It was a rare house in which the valuables were exposed; in a hotel nothing was safe, and what was not stolen was often broken. The nihilism of the intelligent slave, held in bondage in close contact with a society of whites who arrogantly assumed that they were superior, led to unreality and escapism. The subculture of the slave community included not only song and spiritualism, but also a natural sullen refusal to accept reality; fantasy was preferred. Fires were frequent, and nearly always deliberately started by disaffected slaves; the culprit was rarely discovered. After the Civil War, despite all the difficulties, reports of fires in local newspapers were far less frequent than in the boom times of slavery. Violent crimes, common amongst whites, were rarely committed by slaves, who at all times used their native wit to go so far and no further.[33]

There were other habits which failed to commend themselves to the stranger, such as regional addiction to tobacco and corn whisky, of which the first was the more unpleasant. Other people smoked or sniffed the weed – Southerners chewed; they not only chewed, but spat. Perhaps chewing was more acceptable than the communal smoking of the clay pipe, noted in West Virginia in the 1820s, giving both men and women blackened teeth in white faces. 'Such a quaint change', wrote one European, 'from white teeth in black faces.'

Corn whisky had always been not only an article of commerce, but also a form of currency. The attempt to tax whisky as a consumable had led to local revolts against British, federal and state governments; it was often less of a consumable than a store of value, a cash crop. A man from the frontier would set forth from Tennessee, let us say, across the terrible tracks of the period, to visit Tidewater Carolina. The only products of any saleable value he might have had on his farm would be, as quoted in a list of 1812, 'cotton, lumber, pitch, turpentine, tar, furs, animal skins, wheat, peas, potatoes, honey, myrtlewax, tobacco, snake-root, several sorts of gums and medicinal drugs'. What was not listed was corn (maize), which, more than most of these products, was too bulky and too low in value to be worth selling. But the corn could be upvalued and concentrated if turned into whisky, and a packhorse could be loaded with 50 gallons of whisky worth far more than 500 lb weight of anything else except furs and medicinal herbs. The demand for these last two items was too specialist to be universal, while corn whisky would be accepted without question throughout the South. All other forms of wealth for exchange were heavily discounted, except gold and the paper currency of certain states which had a more rigid fiduciary control than was general.[34] Corn whisky of sound average quality was a more reliable form of exchange than any paper, better than anything except gold. But if our friend from Tennessee drank his bankroll and never got to Tidewater Carolina, let alone back to Tennessee, he would not have been the first traveller to have fallen by the wayside. Chronic alcoholism was widespread in a region which could produce the hard stuff so easily, and it was this folk memory which led the Bible Belt to support local option after the Civil War and Prohibition after World War I.[35]

The crippling shortage of gold, reliable paper money or term credit of any stability led to a certain characteristic Southern informality about money matters. The region was chronically short of cash, and had always been so since settlement began; yet economic conditions all tended to encourage the unwise habits of borrowing short and investing or lending long. Perhaps 25–35 per cent of the capital in use in the Southern States was invested in the bodies of slaves. If slaves are to be regarded as livestock (and they were) and if we include horses, mules, cattle, sheep, pigs and chickens, probably more than half the capital in use in the South in 1850 was invested in objects one heartbeat away from death. The hunger for cash, for quick profits, for money-making in general was more evident in the South and southwest than in the North. There were, of course, exceptions. Tobacco plantations in the Tidewater section linked for a hundred years by family connection and business correspondence with London would always have access to money for any serious purpose, and the same could be said of the French connection with New Orleans. In general, however, the appetite for real money, for cheap credit and for a worthwhile form of exchange was never to be satisfied in the South until the present day. Yet much 'money' was in circulation, though some of it was suspect and its origin did not bear investigation.

Sketchy financial arrangements were matched by a high level of violence and much talk of 'honour'. Today we see this portrayed in novels, films and interminable television series, all of which concentrate on the two or three decades after the Civil War. Records are scanty, but a perusal of the newspapers of the period suggests that the rate of death by violence in Georgia, Alabama and Mississippi in the antebellum period was in fact just as high as it was later on. The society lived, of course, on the edge of violence at all times, since half the population was black and often disaffected. The slave-owners were largely first-generation capitalists, determined that the niggers should do the work and be kept in their place; the power of life and death over human beings does not rest well in any soul, let alone the soul of a man whose one aim in life is to escape from back-breaking toil in the beating sun by buying or breeding men to work for him for nothing but their keep.

There may or may not have been thoughtful, civilized men of the stamp or Jefferson or Washington in the Tidewater in the days of King Cotton, but there is no evidence that such creatures ever appeared on the frontier until it was properly settled. It was a rough job to clear the forest, defeat the Indians and hold the slaves, and it was rough men who succeeded and survived. The motto in all disputes was to shoot first and ask questions afterwards, whether the target was white or Indian. Not slaves, of course – you did not shoot slaves; they were property, and therefore flogged – though with care, lest it reduce their value.

The rise of industrial spinning, carding and weaving in England coincided with the first growth of upland cotton in the South and the widespread use of the Whitney gin; as a result, by about 1820 there was an apparently insatiable British demand for raw cotton combined with a huge area available for production in America. For a hundred years Britain would spin more cotton than the whole of the rest of the world put together, and the USA would grow the huge tonnages necessary to keep the mills fully employed. At the end of that period the English Industrial Revolution would be tamed, but one consequence of this excessive growth of mills, in both number and nature, is an imbalance in British society at least as problem-ridden as cotton at the other end of the chain.

If the textile trade in England produced towns wholly populated by wage-slaves, the growing of cotton in the American South produced a society wholly and apparently inevitably dependent upon slavery. There is a certain school of historians in the United States which sets out to prove that slavery was more efficient than free labour. However, an index of the value of a bale of cotton at New Orleans between 1800 and 1850 against the value of a field-hand at auction during the same fifty years shows a fivefold increase in the value of the slave against that of the bale of cotton. The modern historian might reply that the efficiency in the organization of the slave in a gang led to a fivefold increase in his productivity. This may well be true; so is the undoubted fact that the rise in the price of slaves would inevitably lead to improved performance and to changes in the methods of surviving slave-owners, who would otherwise be forced out of business.

This was an era when there was no mechanization, no plant-breeding, no chemical pest control against insects, fungi or weeds, and not even the simplest form of rotational husbandry, manuring or other practices which had been employed for centuries in Europe.[36] It is not entirely possible to reconcile the arguments of the modern historians in favour of the efficiency of cotton-slavery. But there is one factor more significant to an agriculturist than to a historian: there were 500 million acres fit to grow cotton in those states of the South not settled before upland cotton became a crop.

To grow 4 million bales of cotton in 1850 required only about 10 million acres, which included land for growing all the subsistence crops needed for both masters and slaves. This was 2 per cent of the available area. It is therefore probable that, if new land were available, as it generally was, for $1.25 an acre, a cotton plantation-owner could well pick and choose, either leaving fallow, or even allowing to revert to scrub, three-quarters of his acreage, or moving himself every five or ten years to another virgin location. Even if, in an extreme and hypothetical case, an owner grew cotton on just a small part of his holding, say 10 acres, the cost of the interest on that little area of his land would be only a fraction of the cost of the interest in the body of the single slave who would be needed to cultivate it.

Plantation slaves, counting every black from babes in arms to the local equivalent of Uncle Tom, probably stood on the owner's books at about $200 in 1850. The interest on this sum would be at least $10, and in some states up to $15 – in other words, the rate of interest was between 5 and 7½ per cent. On the same basis, the rent of 10 acres of land, valued at from $20 for poor land to $100 for land in very good heart, would be at most only half the cost of the slave. This 10 acres would support all the slaves, plus all the whites on the plantation, and produce the cash crop which kept the whole system going.

There was a very great degree of resilience in the system. If times were good, beef was bought and eaten, but only by the owner; blacks and poor whites, like owners in bad times, invariably ate pork, which was then, in pre-refrigeration times, produced and consumed at home. There was enough timber, usually, to make or repair necessities. There was often brick-earth, stone or wattle clay for building. Iron was recycled. Clothes were homespun, in every sense. Soap was often made at home. Herbal remedies supplied the alternative for every commercial drug – though there were few before 1861 – except quinine. One bad year for the cash crop, cotton, meant some belt-tightening; two years could mean discomfort; three years, hardship. Ruin was much further from the plantation owner in the South than from the Northern merchant or manufacturer or banker. In really bad times the banks would not foreclose, because such action would ruin all the bank's customers.

The informality of money management in the New South also permitted the exercise of clever arbitrage. A loan might be negotiated with a local bank and repaid by agreement when the cotton crop reached New

Orleans. The borrower would borrow local dollars and buy his necessities locally. He would then sell his cotton in New Orleans for gold dollars. He would thus be able to buy up the notes of the local bank from which he had a loan, usually at a discount of 5–7 per cent, and repay the bank with its own devalued currency. Performed often enough such transactions, perfectly legal, much admired, would buy a plantation and all its slaves for next to nothing. The losers were those of the bank's customers who were not so cunning; they were no worse off than trusting lenders in modern times, when inflation is rampant.

Between 1820, when much new land started to become available because the Indians were displaced in large numbers, and 1850 cotton and slavery had formed a fatal partnership. The pattern was fixed and accepted as inevitable by those who lived in the South.

Physical work, it was said, was not for whites. Yet there were many white gangs building roads in fierce heat for money, not because of the lash. Whites and blacks worked alongside one another in river steamboats, in engine rooms whose temperature reached well over 120°F. On the railroads, many gangs included whites and free blacks. Only the cottonocracy maintained that black slaves were necessary to work in the fields because of the fierce heat. The lie to this myth was obvious by 1872, the year in which the South produced as much cotton as in the last year before the Civil War. The majority of it was produced by sharecroppers, most of whom were white, not black.[37] But most Southerners still prefer the myth.

There was a huge problem if peaceful emancipation were to be achieved, say, in the 1850s. The value of the slaves in 1850 may be safely and conservatively estimated at over 2 billion 1850 gold dollars, or about $50–60 billion in today's dollars. It was not much – less than the value of General Motors, IBM or Exxon today – but it was a vast sum in terms of the United States economy in 1850, about ten times as large as the federal budget, about a quarter of the gross capital value of everything in the slave states, and about ten times the value of the cotton crop itself. The South had become capital-intensive, the capital being the value of the slaves.

Before cotton became so important, three attempts at emancipation had been made. The first, already alluded to, began in the South in the 1780s and foundered on the compensation issue. So did the efforts in the 1820s, and those in 1834 following the example of the British in the West Indies. By 1850, the compensation issue had gone from difficult to impossible.[38]

It is one of the awkward facts of history that obscurantist, backward, tsarist Russia emancipated the serfs two years before the free, progressive, democratic United States emancipated the slaves. Yet the reason is obvious after a moment's thought. Serfdom tied men, women and children to land, the working of which was exchanged for some days' labour each week on the lord's estate. There was the lord's land, worked exclusively for his benefit, and there was the commune land, worked by the serfs for their sustenance. The number of days to be worked for the lord and for oneself varied from country to country, district to district, crop to crop and century to century. The system was very stable and had a long and respectable history – more than a thousand years in some parts of Europe, and more than two thousand in parts of Asia. It was one of the disadvantages from which white Europeans escaped to America, yet it was far more humane than slavery. Such was the bad name of feudalism that the Founding Fathers preferred to perpetuate slavery rather than convert it into a form of feudalism – already at work in the case of white indentured servants, who laboured without pay for seven years or so and then gained a land grant. At Independence the system was not thought applicable to blacks; yet if, then or at any other time, the slaves had been turned into serfs, slavery as an issue would have been destroyed and the Civil War never fought. While there is plenty of evidence that a form of feudalism for blacks had been considered in the 1770s and 1780s, there is no record of this compromise method of emancipation having been discussed in the 1840s and 1850s.

Such a method, involving free labour by the ex-slave for seven years and little cash compensation, was tried in the British West Indies in the late 1830s. Unfortunately, on many islands the ex-slaves ran away from the sugar estates and squatted on empty land, refusing work, growing their own subsistence crops and withdrawing from economic activity. Many sugar estates were ruined, without labour of any kind at times, and put into terminal decline. It may have been this bad example which vitiated any attempt at emancipation with compensation in the American South. Perhaps such a method was more suited to a country where land was scarce and labour was plentiful. Certainly the system worked in those West Indian islands which were densely populated.

No one in the United States could foresee an end to slavery which did not follow one of four roads: first, 'withering on the vine' – in theory, some people held, slavery was going to wither away, as inefficient in the USA as it had proved to be in the West Indies; but this theory ignored cotton's monopoly position. The second option included compensation, which was considered impossible. The third, emancipation without compensation, would have ruined the South; while the fourth, violence in the shape of war, is what actually happened.

Faced with the apparent inevitability of slavery, apologists claimed that paternal slavery was a kinder relationship than the hazards of the boss/worker syndrome. Blacks were treated as children from the cradle to the grave, while for the Northern worker there was no social security, no job protection, no health insurance and no employment pay. Freedom involved the freedom to starve, to be unemployed, to be sick without help, to face old age with insufficient means, but it also involved the spirit, and the options available to free men. For the black slave there were no options – not even, on some plantations, about his food, which may well have been cooked in a communal kitchen. The peculiar rules relating to slaves, such as the denial of all leisure except on Sunday; the enforced illiteracy; the illegality of manumission in some states; the 'pass' laws; the slave patrols and the hated bloodhounds; the professional slave-whippers; the overseer, usually 'po' white trash'; the curfew in many cities; the discouragement of marriage and the splitting of mothers and children; above all, the fugitive slave laws and the treatment of absconding slaves; all these cruel features were peculiar to the Peculiar Institution, not because slaves were slaves, but because slaves were blacks, and blacks were different.[39]

Whether the harsh laws were economically necessary or not, their existence very greatly increased the moral majority against slavery. Most members of this party had, like any other moral majority, good self-interested reasons which were first politicized and then elevated and sanctified as moral issues. The poor whites in the South were not in favour of slavery; this was proved in West Virginia in 1861 and would have been proved in the rest of the Appalachian mountain region if whites had been offered the choice. But they were not allowed to vote, so they rode and marched and fought and were starved, wounded, killed or ruined for a cause in which they did not believe, from which they did not benefit, but which they did not question.[40]

In the North, self-interest joined with morality to confirm the workers' view of slavery, but manufacturers, bankers and men of commerce did not necessarily agree. Skilled craftsmen have always looked down on the hewers of wood and drawers of water.[41] This attitude was not helpful in resisting the detrimental effects of the Industrial Revolution. It was the aim of every man, however humble, to work himself out of the necessity to earn his living by mere sweat.

The problem of organizing a workforce has always troubled all human societies. It is a difficulty which is exacerbated by the modern Western tendency to replace people by machines, allied to the idea of giving people the expectation that they have control of their own destiny, which is loosely called 'freedom'.

Contemporary socialist countries are much better placed, in theory, to organize the workforce by 'planning' because Marx believed that the State had the right to the product of a man's body, and most socialists accept this dictum, as declared at Gotha in 1875. The concept does not differ in essence from that of the slave-owner having the right to the slave's work for life, except, of course, that the State is held to be less wicked than the kindest slave-owner.

The origin of one man's right to another man's work is primeval. Two animals meet in combat; the loser loses everything. In most animal societies, the loser also loses his life. If the loser is edible, and the winner is carnivorous, the loser is eaten. If the loser is female and of the same species, she is taken for the male winner's use. If the loser is male, and has female(s) in tow who may or may not be the object of the male rivalry, the female(s) become the property of the winning male.

Except in the details, mankind is not very different. Male losers' lives were spared in exchange for lifelong slavery. These were the earliest forms of slaves, far older than bought or bred slaves. Most men would prefer slavery to death, except in poetic times of high emotion. [42]

Until the nineteenth century it was not considered excessively inhuman to put prisoners of war to death if they could not be safely held in custody. The PoW's right to life was not assumed by the Japanese in World War II until it became likely that the Allies would win the war. The right to the unrewarded work of the prisoner was accepted by all European societies as late as the eighteenth century. An alternative was ransom: cash for a lifetime's work. [43] Once this is accepted, slavery becomes reputable and much else follows. Feudalism is preferred to slavery, and the Marxist position becomes respectable.

Modern Western society demands the best of both worlds. We want full employment and absolute freedom to choose and change our jobs, as well as the right to retain any profits or savings we may make. We also want the State to put a floor under our indigence, to provide free education for our children, to care for us in sickness and old age. None of these rights or freedoms existed in the United States or Europe before 1861, or indeed for many years afterwards.

Increasing mechanization has also led not only to political demands for these rights and freedoms, but also to the question of leisure, a problem for the intellectual and an obsession for the unenterprising. The capacity of most people to occupy themselves usefully without spending money is diminishing.

Slaves cannot have leisure, since it does not pay the owner to let them do so. The wage-slave has the minimum leisure that the wicked capitalist has, unavoidably, to concede. The true amount of leisure that people would actually like to have is unknown, variable and subject to individual choice.

Until relatively recently the majority of the population in Europe or America had to sweat to earn a living, yet the manual worker could be despised or pitied as a poor unfortunate being who was assumed to have no education, no native skills and no aptitude for the better things in life. The nobility of manual toil, that mid-Victorian concept much promoted by Ruskin, did not appeal to those who knew what work was. The poorest white in America would trade, gamble, bargain, cheat, lie or steal, according to his conscience, to escape physical work, like the poor anywhere in the world. The way to escape was to become first self-employed and then a capitalist: in the American North, to hire others to do the sweating; and in the South, to buy their work for a lifetime.

There was, the Southerner said, no real moral difference between employing cheap Irishmen off the boat to dig drains in Boston and letting them go when the job came to an end, and buying or rearing slaves to dig ditches in Georgia. But since the commitment in the South was for life, the slave-owner was of a higher morality than the construction boss back East. This was the welfare state against freedom in its most acute modern form: protagonists of both sides were as irreconcilable as the exponents of socialism or conservatism.

The real argument against slavery was not economic, nor moral, nor religious, nor social. Economists and moral philosophers argued the case for both sides. With the exception of the Roman Catholics the Churches of America, which had once universally accepted the slave trade, had by the 1820s and 1830s split between North and South on these issues. By 1850 an Episcopalian, a Baptist or a Methodist could find his Church conveniently supporting slavery in the South and, just as worldly-wise, condemning the practice in the North and havering in the border states. The social position of the slave could be represented as being better than that of the poor immigrant worker, often unemployed in the North, or the cotton operatives of Lancashire who were worked far harder for less reward than most slaves.

The real argument, particularly against slavery in America, was that of expectations. Most whites in America had dreams for themselves or their children of becoming farmers, landowners, great merchants, lawyers or statesmen. This was the land of opportunity. Most emigrants to America had come in search of the better life in this world, not the next; everyone had a chance to get out of the harsh present into a golden future. From this it was a short step to proclaim that if a man could not get out of his unpleasant present situation, there was something wrong with the man himself.

The slave was the exception. No one could blame a slave for not improving himself, for any increase in skill, productivity or know-how would accrue only to the master. The slave had no hopes either for himself or for his children, and he was not encouraged to seek the consolation of religion to make good the deficiencies of this world.

Throughout history the passivity of the slave has always led to the decay of the system which exploits this means of servitude. Any argument that slavery is a more efficient means of organizing labour than either the feudal system or the wage sytem is only true if the slave produces as much surplus as the serf and as much as the worker who toils for cash reward. But anyone who has tried to make a living out of any kind of stock farming knows that the relationship only yields maximum profits if the animal is healthy, happy, and has a zest for life. In the South the 'stock' and the beasts of burden were slaves. Whereas an ordinary small Southern farmer would work alongside his slaves, cajoling, joking, inspiring them to optimum performance, plantation cotton slave-owners had little contact with either animals or slaves – they grew crops. They would have quality riding-horses, but these would be treated like an expensive fast car, their equivalent today. The main relationship of the stock – the black slave – was with other black slaves and with the overseer, usually an unpleasant man despised by owner and slave alike. So husbandry never entered the relationship.

As far as incentive was concerned, the relationship must have been in essence short on carrot and long on stick, and inherently unstable. In all successful systems using rewards, the person rewarded must feel that by his conduct he can affect the degree and extent of the reward. The slave must have some feeling that if he worked harder, he would gain some advantage. What were the rewards? Special meals, parties, dressing-up, competitions and dances like an open day at an infants' school – did these constitute incentives to grown men and women? In fact, of course, most slaves led a dull, boring existence enlivened only by more or less work, by an assumption of passivity which fooled overseers, owners and politicians alike. This passivity led, as always in slave societies, to complacent self-satisfaction in the owners; to over-confidence about a system which was socially unstable as well as morally indefensible; and, above all, to a way of life which could only change by means of violence.

Rich Southerners had plenty of time to think about the future of their region. Most of them spent their ample leisure in other pursuits: drinking, gambling, riding, going to parties, travelling, trading, visiting relations, gossiping – all the attributes of an elegant new Greece or Rome were available to a minority of the rich. But neither Rome nor Greece had had as neighbour an industrial society based on cheap coal, cheap iron ore, cheap transport, and, in many ways, the most ingenious managers in the Western world. It is possible that if the cotton kingdom had been an island, isolated, threatened by no other power, the cotton–slave syndrome might have continued. Contiguous with the Northern states, economically challenged, socially surpassed, industrially overtaken by the Yankees, its leadership lost to those whom Southerners regarded as vulgarians, the cotton kingdom refused, like an alcoholic or drug addict, to contemplate the future.

Meanwhile in the North, in good times or bad, men of all classes lived in hope. Manifest destiny was the birthright of every Yankee. One of those who dug drains in Boston in the 1850s was the ancestor of the American Kennedy clan. By what means we do not know (though we can guess), he very soon became his own boss, and then the boss of others, and then a politician.

The black who was digging ditches in Georgia at the same time would need the bloodiest war fought by whites in the nineteenth century to make him 'free'. Nearly one million whites were killed, wounded, or suffered in some personal way in order to free (or not to free) 4 million blacks. When it came, that freedom was of base coin.

It would be three generations before the blacks of America came to share the expectations which whites regarded as theirs by right. By the time the blacks had achieved some kind of rough equality with the poorer, less able whites, the expectations had ceased to be physical, the West was populated, and the white drain-digger's great-grandson was in the White House, talking about the 'New Frontier'.

This seemed a cheap marketing concept to many blacks; first came freedom with poverty, then for the next generation freedom without education, and for the next generation after that freedom without equality of opportunity. Few blacks ever remembered, in either 1850 or 1960, that at all times the slave or ex-slave in the South was better off than his relations in Africa. This argument has never appealed, and it was difficult for the slave to accept his lot when the whites around him lived in a land which offered more opportunity to more individuals in the nineteenth century than probably the whole of the rest of the structured, caste-ridden Old World put together.

It was opportunity that the blacks missed, but they called it 'freedom'; when they got freedom they found that it did not contain opportunity. This terrible disappointment has laid up for America a harsh racial problem, possibly for all time, with the less able blacks at the bottom of the heap, and without the opportunities to lift themselves out of the mud which were open to all whites in the nineteenth century. This was the echo of the firebell in the night – the legacy of slavery. It was the long-term influence of cotton upon the United States.

When the American Civil War came in 1861, neither side wasted much time in asking why it had broken out. Wars require for their inception euphoria, adrenalin, a form of madness. At least a dozen reasons are given for the origins of the War between the States, and at least five can be selected and considered separately and in combination. Among those reasons outside the mainstream is the explanation given by Mark Twain, who blamed Sir Walter Scott. By this he meant that a region of the United States with a low rate of literacy, and which read Scott more widely than anything except the Bible, was likely to favour the romantic view of the chivalrous white man, defender of the True Grail and of the honour of Southern womanhood. Significantly, Northerners preferred Dickens to Scott.

If Scott was widely read in the South, he was not read intelligently. Beneath the floss of medieval standards lie nuggets of realism which Southerners probably skipped. A hard, sober view of their situation would not have led to violent secession and civil war, however often it had b~en threatened in the past. Southern perception did not include the possibility of being defeated by the North, a region which commanded many times the resources and income of the Confederacy. Lack of realism characterized many other calculations ('One Southerner can whip five Yankees') and was a direct result of the superiority conferred on the whites by slavery and by the natural contempt felt by an agrarian civilization for a mongrel industrial proletariat. In the end, of course, the hubris of the Southerners needed nearly four years of war to effect the inevitable, and war-weariness in the North made Southern defeat seem not always inevitable. The War between the States was in many ways the American *Iliad*, more important to true nationhood than was the War of Independence; it was the fire which forged the Union.[44]

The North could not believe that the South would fight, and when the Confederacy became organized Northerners found it incredible that all the white males would fight for a system from which the majority did not benefit. Only half of all Southern families owned any slaves; only 10 per cent more than one; only 1 per cent more than five; only one in a thousand families more than fifty. On the other hand, poor white trash, however poor, however illiterate, despised and pitied their opposite numbers in the North and believed in King Cotton. Many of the poor whites also believed that one day they too might become slave-owners. Many others were carried unthinking upon a rhetorical tide of froth and spume, with much talk of states' rights and freedom. The slaves remained impassive, silent spectators of a quarrel which was really about their status, though for two years politicians pretended otherwise.

In the first 'modern' war in history the Southerners failed to calculate the material disadvantages of their position. Although there had been thirty years of cold war, they were totally unprepared for the reality. There were only short stretches of railroad in the South; most goods went by water. There was not a Southern city of any size through which heavy trains could pass; nearly everything had to be trans-shipped. There was no workshop in 1860 which could rebuild a locomotive, let alone build a new one. There were no carriage or wagon works; no rolling mills for rails; no glassworks for windows. Of the 1860 cotton crop, consisting of over 4 million bales, nearly 80 per cent went to Europe. Cotton was one of the bulkiest raw materials, and certainly the most valuable, in the whole United States. Out of the 700,000 bales used internally in the United States, only 2 per cent went more than 50 miles by railroad; 98 per cent was moved by coastal shipping. Though the railroad network in the USA as a whole exceeded that of England in 1860, in the South most of the traffic was local and 90 per cent of revenue came from passengers.

In other respects too, material conditions in the South did not favour a long war. There was no woollen mill; no linen mill for tents; no shoe factory; no manufacture of decent paper, ink, lead pencils, matches, pins, needles or watches; no pharmaceutical manufacture, not even of the essential quinine. All these things had previously been imported from Europe or the North. The South even imported salt; so much for self-sufficiency.

Of all these weaknesses, shortage of transport was the worst, for it led to shortages of food in the army (with food rotting on farms), to shortage of ammunition, uniforms, boots and even wagons. After the first few months of the war, the once magnificent Confederate Army looked like a host of scarecrows.

So convinced were the Southerners of the power of King Cotton that they had acted with incredible naïvete. Instead of selling the 1860 cotton crop for gold, the South withheld supplies from Britain on the grounds that within a few months the British, desperate for cotton, would break the inevitable blockade and recognize the Confederacy. Nothing could have been further from the truth. At the beginning of the Civil War there was a world surplus of cotton at all stages of its conversion.

In 1862 one-third of all spindles in Lancashire were idle and another third on short time; a year's supply of raw cotton and half a year's supply of cotton greycloth were waiting in store in Liverpool and Manchester.[45] More unemployment was due to the trade cycle than to the Civil War. One mill-owner made more money in the first year of the war by buying and shipping raw cotton from Lancashire to New York than he had made in the previous twenty. Why were the Southerners so foolish? The truth could be read in the London *Economist* in 1859, 1860 and the spring of 1861. One wonders how many people in the South read the *Economist* and, if there had been more, would there have been a Confederacy?[46]

After the beginning of the blockade, shortages and inflation began to develop in the South as soon as the spring of 1862, and Confederate leaders regretted their short-sightedness. The brokers, spinners and weavers of

Lancashire were encouraging alternative supplies, mostly from Egypt and India. Cotton piled up in warehouses all over the Southern States.

There was no commensurate support from the majority of the European cotton industry. Opinion was divided. Most politicians favoured neutrality. English cotton bosses, representing an industry which produced a third of all exports and a tenth of all economic activity, sympathized with the South, and at one time in 1862 recognition of the Confederacy looked possible; Gladstone, the future Liberal Prime Minister, was one of those who supported recognition. The poorer classes, taking their cue not from unemployment in Liverpool and Lancashire, but from the strong moral case preached by nonconformist ministers and radical politicians, were staunch supporters of the abolitionist cause; Abraham Lincoln, despite never having visited Europe, became a cult hero. Such feelings were reciprocated; Lincoln described Lancashire's support of the North against shallow self-interest as 'sublime Christian heroism'. It was indeed: in 1862–3 more than three-quarters of the mill-workers of Lancashire were suffering hardship.[47]

It is tempting to castigate the leadership of the South for their inability to foresee the inevitable; to wish that more time had been spent on working out methods of growing cotton which did not involve slavery; to hope that mankind is not always so foolish as to need violence to effect change; to wish for evolution rather than revolution. Plenty of people in today's world are so conditioned that they cannot acknowledge their own obsolescence, and are so defensive about their own present state that they regard everything about it with a kind of aggressive complacency. A quick mental trip round the world produces at least half a dozen 'Confederacies', alive and well and as dependent upon some unsustainable factor as the original Southern Confederacy was upon slavery.[48] In each case, as in that Confederacy, as long as the original premise is accepted, the argument is logical and unanswerable.

In 1784, before cotton became a Southern crop, the whites in the region had the highest per capita income in the whole USA, more than 50 per cent greater than the ex-colonies as a whole, and nearly 75 per cent higher than Britain. Cotton prolonged this superiority into the Industrial Age. Yet the methodology of cotton-growing was an anachronism. The cottonocracy was a pre-industrial agrarian civilization, not needing or using much iron, steam or mechanized transport. Given the invention of the gin, and given the land, it could have existed in Greek or Roman or medieval times. In nineteenth-century America, slaves were considered essential to cotton.

This proved to be a fallacy. Within seven years of the end of the Civil War the output of cotton had reached prewar levels, using no slaves and hardly any free black labour. Given the need to work, whites in the antebellum South, aided by share-cropping and abetted by mechanization, could have produced as much cotton without any slavery at all. How many of the survivors ruefully realized this truth in the 1870s? If the whites had known how quickly the devastated, bankrupt South could recover after the Civil War, and how cotton could be produced without slavery, it is very unlikely that anyone would have allowed King Cotton to drive them into a war which they could not logically have ever hoped to win.

Was this, then, the last great agrarian civilization? It lasted a few decades, fuelled by overseas capital, and with 80 per cent of its product going overseas. Yet it did not consider itself a 'colonial' economy. The grandparents of the men who led the South in 1861 had led the nation against the imperial power. Like few ex-colonies, the young United States was richer than the Mother Country, and cotton prolonged these riches in an area which was in other respects being left behind.

The greatest irony of all is, of course, that this great agrarian slavocracy depended upon the steam and iron of Europe and New England for its market. The last great slave empire fed the first great industrial revolution. Each was dependent upon the other, in a symbiotic relationship.

All over the world, cotton textile manufacture became the first element in the first Industrial Revolution. Textiles in general, and cotton in particular, have always been the first processes to which men have applied, in rational form, the power of water, steam and then electricity. The first great manufacturers were those of cotton, whether in England, France, the rest of Europe, the USA or in any emerging country in the nineteenth and twentieth centuries.

In 1861 cotton was by far the most important industry in the USA: $55 million was added to the value of raw cotton, and the industry had 803 cotton mills, each employing on average 143 people. Few were in the Southern States. In the decade before the Civil War the American industry took less than 25 per cent of the crop, the French about 11 per cent and the rest of the world, barring Britain, about 6 per cent. This left nearly 60 per cent of the American crop to be processed in England, mostly in Lancashire, which also processed nearly as much non-American cotton as the rest of the world used.

So it is right to talk in terms of America and Britain. America produced two-thirds of all raw cotton exported throughout the world, and Britain exported more than two-thirds of all manufactured cotton products. Given a couple of years, the Southern States and Lancashire could between them have made good any loss of production in the rest of the world. In these two countries, Britain and America, cotton changed history.

In 1784, the year the first bale arrived at Liverpool, there was virtually no cotton grown or manufactured in the United States. By 1861 cotton had become the single most important crop traded in the world, and more than 80 per cent of that crop was grown in the USA. In 1784 there was not a bale of cotton processed by steampower. By 1861 the skies above Lancashire were black with smoke from steam-raising boilers, whose power was devoted almost exclusively to the cotton industry.

In 1784 there were fewer than half a million slaves in the thirteen colonies. By 1861 there were nearly 4 million, and slavery was an issue which had led to the bloodiest and most expensive conflict of the nineteenth century, much more costly than the Napoleonic Wars before 1815 or the Boer War which started in 1899. In 1784 the thirteen colonies were populated by white men of very similar stamp. By 1861, while the whites in the South remained recognizable descendants of the men of 1775, those in the North had all the trappings of division conferred by industry and occupation.

In 1784 there was a handful of spinning mills in England; most workers were adult males. By 1861, female and child labour in Lancashire had become a disgrace which aroused the indignation of all humane people. In 1784 there was no Cotton Exchange, no real infrastructure, no means of public investment in the textile trades. By 1861 all these sophisticated economic advantages existed in New Orleans, Liverpool and Manchester.

In eighty-five years cotton had become the norm, in quality, price and delivery, against which all other thread and cloth was judged. In 1861 the bulk of the raw material was grown and prepared by slaves, carded by machine, spun by steam-driven devices tended by children, and bought and sold in modern markets.

In the process, the cotton states had become fixed in a way of life for which no one could predict a peaceful future. More than half of the working population consisted of slaves of a different race, and the whole economy appeared to depend upon the euphemistically named 'Peculiar Institution',[49] an aberration. To be cured of that aberration the South endured war, reconstruction and inferiority.

In its wake the Civil War left the most difficult of all American social problems, that of the relationship between black and white. In the 1960s Southerners said that it would be the South where integration would first prove genuinely possible. Southerners today remember that the first extensive city race riots took place in the North.[50]

Compared to the American 'sunbelt' or to Europe, modern urban England is no object of bright pride. Liverpool and its environs came to be responsible for the import of most of the raw cotton shipped throughout the world, and for most of the export of the manufactured product. Today, there is hardly a cargo ship to be seen. A sorry tale can be told of bad economic planning, terrible high-rise blocks destroyed by their inhabitants, wholly inappropriate education, corrupt and obscurantist trade unions, dishonest and short-sighted politicians, a city urbanized very quickly and very badly and with no subsequent organic change: the history of decline is inexorable. Today, Liverpool is a great stranded whale, without function, a purpose-built city whose purpose has disappeared, an expensive mess which in 1983 cost the rest of the UK £2000 per Liverpudlian to maintain. Liverpool's chief claim to the notice of the civilized world is its vigorous football sub-culture, which regrettably includes a number of White Yobs, the most graceless underclass in Europe.

Such a process is not inevitable, and Liverpool is not Lancashire. Smaller places – Burnley, Rochdale, Nelson – were built as single-purpose towns before 1850. A century later, without the curse of any 'planner', they had begun to diversify into engineering, new techniques, service industries. One old cotton mill even became the world's largest deep-litter chicken complex outside the USA.

No matter that the changes must be repeated again and again. Some communities have learnt to change rather than fear the future; to adapt rather than regret the past; to regard evolution as inevitable; to encourage talent, not to drive it away; such communities *live*. Conversely, if it cannot adapt, a community – family, tribe, village, town, city or country – will fade out like any other species overtaken in a Darwinian world.

Notes

1   Rathbone MSS, University of Liverpool. The boy's name was Compton.
2   The 'Liverpool Party' was the name given to the ambitious leaders of that port. By 1800 they had successfully supplanted Bristol and London as the chief home bases for the triangular slave trade, involving rum and sugar as imports. Their position in relation to Lancashire's cotton, to Ireland, and to the iron, coal and metal works of the Black Country made their aim of becoming the most important trading complex in the UK a not impossible one. In the end, the Liverpool Party exerted great influence on capital markets and was instrumental in the development of turnpike roads, railways and canals, but never became more than a powerful second city to London.
3   By any standard, cotton was the most expensive pre-industrial thread, cloth or clothing – more expensive than silk or fine wool.
4   'Staple' in terms of textiles means length of fibre. The long-staple cotton imparts more continuity to the spun thread, as does the built-in 'twist' of the fibre. The hollow nature of the fibre gives cotton an advantage over man-made substitutes – the trapped air offers better insulation, greater comfort, and more 'feel'.
5   Different varieties of the same botanical species can be found in the tropics, which in summer have a twelve-hour day or thereabouts, and in high latitudes where the midsummer day may last for twenty hours or more. Some plants, such as grasses, adapt easily; others do not. Cotton sometimes adapts and sometimes cross-breeds, which is adaptation by another means.

6 On primitive looms, if a rectangle of cloth is required it is of little importance whether it is the warp or the weft that faces the weaver. But if a length (or bolt) of cloth, say 3 feet wide, is required as the end product, the warp must run the length of the cloth. The thread must also be long, and at least twice the strength of the thread used in the weft. Perhaps it is because men's arms tend to be longer and stronger than those of women that they traditionally undertook the weaving.

7 *Encyclopedia Britannica*, 3rd edition, 1788.

8 The theory of division of labour must have occurred many times in history to any individual who had to make, say, twenty objects each requiring several processes. Thus twenty scones can be made from the same mix, baked together, and dusted with flour together. To repeat these processes each time for each scone would seem bizarre. From this division of labour by time, it is only a short philosophical step to divide by function. Thus two, three or four people might each be performing one process in a hypothetical scone factory. From dividing by function to mechanizing by function, say mixing the raw materials for the scones, it is another short step. This is what happened in the cotton industry, and in every other. After the publication of Adam Smith's *Wealth of Nations* in 1776, the theory of division of labour became axiomatic and so obvious as not to require exposition.

9 The great economic advantage that Great Britain had over France from the *ancien régime* until after 1815 can be shown by the fact that though there were over a hundred water- or animal-powered cotton mills in the UK in 1790, there were only two in France, both of which were owned (and managed?) by the English. This industrial head start helped the UK to defeat France and her Napoleonic satellites, who outnumbered the population of the UK by four to one. (See Arthur Young's *Travels in France, 1787–9.*)

10 'Grave clothes' were not only worn by the mourners, but were also specially designed and made for the corpse.

11 To rove means to draw out, lengthen and slightly twist the yarn. To card means to comb out the thread, raise the nap and make it naturally parallel.

12 It is worth considering what would have happened if all the raw cotton imports had continued to come from all over the world after 1800, rather than more than 90 per cent from the United States; or if Ireland, with its reservoir of unskilled workers, had been to the south of England and not to the west; or if coal had been available in the hinterland of Southampton, rather than Liverpool.

13 What the comparison between the slave in Georgia and the poorly paid worker in Lancashire failed to mention was that the inherent advantages of space, climate and resources available in Georgia were not mirrored in Lancashire, with its damp, cold weather, overcrowding, and houses built of local materials – brick, tile and slate – which are all poor insulators, making buildings hot in summer and cold in winter.

14 The Enclosure movement was originally designed to make better use of the common fields of medieval times, and allotted to each commoner a proportionate area of land or cash in exchange. The concept, once beneficial and economically efficient, became, according to the best evidence, a vehicle for the lord of the manor to enrich himself at the expense of those much poorer than himself. On the way the lord (who might be titleless, but was certainly rich) also acquired power – often voting power in parliamentary elections – as well as a landless class of peasantry to till his fields, and a reputation, deserved or not, as an exploiter of the people. What undoubtedly happened was that the landless peasantry bred beyond the ability of the diminished rural resources to support them, and their children became the new proletariat of the towns. See also the chapter on the Potato.

15 A 'spinster' could be replaced by a machine spindle costing less than one month's wages in 1784. Today a robot to do one job, such as tightening up a single screw on a production line, costs say $5000. To achieve the equivalent of what a machine spindle did (four functions), today's robot in, say, welding, would cost considerably more than a year's wages.

16 The Whitney gin only answered the problem of the short-staple upland cotton, and was a substitute for the age-old roller gin of Asia called a *churka* in India and a *manganello* in Italy and Spain. Whitney's invention had an output fifty times that of hand-picking, whereas that of the *churka* was only five times the output of the hand method.

For Sea Island, long-staple cotton the Macarthy gin was developed about a hundred years ago. The long fibres, which would be damaged by the saw teeth of the Whitney gin, are drawn by a leather roller between a metal plate called the doctor and a blade known as the beater, which forces out the seeds. In the late nineteenth century the best material for covering the rollers seemed to be the stomach hide of the sea walrus – 25,000 animals were slaughtered every year to meet the requirements of the cotton industry, and their carcasses were often left to rot.

17 The factory was built in Connecticut, rather than in any cotton-growing state, not only because Eli Whitney was a Yankee but because it was almost impossible to obtain in the South a supply of free, white, skilled local workers. This was an early indication of what would prove to be the greatest Southern deficiency when cotton led to war.

18 The English demand for cotton grew as follows, and the cost of yarn, expressed as for export, fell (table from the US Bureau of the Census).

	Annual consumption (lb million)	% from USA	Cost of yarn, spun in England, in US c/lb, f.o.b. Liverpool
1810	79	48	145
1820	129	54	90
1830	248	70	60–70
1840	350	79	55–105 (1835–45)
1850	588	86	35–75 (1845–55)
1860	1083	92	28–55 (fluctuation during 1860)

Yarn fluctuated less in value than did raw cotton.

19 There are various permutations of the expression 'The South'.

The South means the area south of the Mason-Dixon line, fixed in 1763–7 to settle a dispute between Pennsylvania and Maryland. The line was extended westwards and formed a boundary north of Virginia, Kentucky, Missouri and Texas.

The Old South consists of the areas east of the Appalachians which were important at the time of the War of Independence: Virginia, North and South Carolina, Maryland and (sometimes) Delaware.

The Deep South is really synonymous with the cotton kingdom, or the plantation economy, or the slavocracy. It consists of parts of inland North and South Carolina, Georgia, Alabama, inland Louisiana, Arkansas, and the cotton-growing area of Texas.

The New South, or the southwest, is a moving target. In 1800 it comprised southwest Georgia and northern Florida. In 1860 it covered only Texas and Arkansas. A straight line drawn from Washington to Dallas would mark the central axis of the southwest at any time.

20 Importation of slaves into the United States became illegal in 1808, but continued either by means of smuggling, or via Texas, which was first Mexican, then became independent in 1836, and finally American in 1845. Virginia and other worn-out tobacco areas found slaves more profitable than any other product. For a sad picture of the faded Virginia gentry between 1784 and 1880, after which date slave-breeding came into its own, see Jan Lewis, *The Pursuit of Happiness: Family and Values in Jefferson's Virginia*, Columbia University Press, New York, 1983.

'Virginia' tobacco only became important after the invention of automatic cigarette-making machinery, a century after the Revolution. Between 1774 and 1875 Virginia tobacco, grown also in Kentucky and North Carolina, was of no great account. Though it represented one-sixth of all US exports by value in 1774, it fell to much smaller proportions in the nineteenth century. See Lord Sheffield's *Observations on the Nature of Trade between Britain and North America*, 1783.

21 Tidewater means the area washed by salt water, from Delaware in the north to Georgia in the south, but it originally indicated only the Chesapeake Bay region. The term usually included land watered by rivers to which ocean-going ships could sail without natural hindrance, i.e. the Savannah River, navigable by sailing ships for at least 100 miles, or the James or Rappahannock rivers in Virginia. The Tidewater settlers thought of themselves as gentry as early as the 1670s, at which date tobacco was already in surplus due to over-production, the land was half-exhausted, and the aristocracy in debt to English bankers and factors.

22 Piedmont is the area below the Blue Ridge Mountains in Virginia, and its equivalent further south and west – level, well-drained, well-timbered country where an able man could live out a subsistence life, making his own clothing and his own whisky, forging his own bullets out of local lead ore and iron from the soil. Tobacco, corn whisky, skins, furs and medicinal roots would be the cash products. The only import needed from England or from the rest of the United States was gunpowder, plus really good steel for axe-heads. The Virginia part of this area was already nearly fully populated by 1750, and people moved into the Shenandoah Valley long before the Revolution. Further south, the Piedmont region became the home of cotton in the Carolinas and Georgia.

23 Amerindians in what is now the United States were thought, in days gone by, to number as few as 3–4 million across the whole continent. Recently the Indian cause has become a very fashionable and emotive issue, and exaggerated estimates as high as 50–60 million have been promoted. No real case can be made for either of these extreme figures, but an upper limit of 25 million seems plausible.

24 The prime resource of all people in the world – except of a few cultures such as the Australian aborigines and the Eskimos – has always been land. The distinctive feature of the 'new' empty countries colonized by Europeans was that there was much more land than there were people, the reverse of the European position. Thus in time the creed of 'Waste not, want not' was abandoned, and reached its nadir in the throwaway culture of the 1960s. The limitation to the more grotesque forms of conspicuous consumption is no longer land, but more probably energy.

25 In 1803 Napoleon, anxious to clear the decks to renew the war with England, reversed his policy on the Mississippi Valley. He and his Foreign Minister, Talleyrand, had earlier dreamt up a grandiose scheme for cutting off the westward progress of the United States by uniting New Orleans with Canada by means of a string of French settlements west of the great river. In essence this was the same plan as that of the 1750s, at the time of the Seven Years' War, writ large. In a rapid *volte face* 'Louisiana' was now sold to the young United States for $15 million; the area concerned comprised not only present-day Louisiana, but all or most of twelve other future states. It represented a good deal for both sides. Napoleon was not strong enough to carry out his own ambitions, and French troops were in more trouble than they could handle in what is now Haiti.

26   The North American colonies were the traditional suppliers of meat, grain, dried fish and butter to the West Indies. In 1770 'grain' included wheat, barley, oats, peas and beans, bread and flour, potatoes and rice, as well as 'Indian corn' or maize. (See Lord Sheffield's *Observations*, 1783.) After Independence, much illegal trade went on between the new United States and the British colonies. It is doubtful, however, if the trade after 1784 included, as Lord Sheffield tells us it did in 1770, 163 complete timber-framed houses.

27   Rotation of crops evolved in the Old World only because of high land cost, in turn due to density of population. The reverse condition, of cheap land and a low density of people, also implied a shortage of workers in rural areas. In the USA this led, in the North, to intense mechanical development, e.g. ploughs, discs, reapers, binders and combines; but in the South there was only an increased demand for slaves. Jefferson notwithstanding, it was not bad husbandry which made Americans 'bad farmers', but business sense.

28   Sectionalism is not dead yet. Compare Washington DC under Johnson's Texans or Carter's Georgians or Reagan's Californians. The style and content of the nation's capital reflects the origin, companions and prejudices of the President as much today as ever.

29   Even now, there are traces of settlement in the emptier parts of Virginia and the Carolinas which show that a farm was abandoned as long ago as the 1780s. The same was even more true in the less populated times of the Civil War, when huge numbers of soldiers of both sides invaded uninhabited areas and came across one-time habitations dating back to the 1660s.

30   We have no idea in today's power-assisted world how in all previous centuries every possible device has been employed to avoid physical effort at work. If wind, water or animal was not available, then the fortunate found a servant, serf or slave to relieve the master from the tedium of toil. Steam removed the unskilled muscle requirement from many aspects of manufacturing, transport and mining, but was of little assistance in agriculture. Electricity and the internal combustion engine have done more to liberate men and women than any previous technical advance, and have been much more influential than any ideological movement or political party. If it were not for these forms of power, the position of the proletariat in a dictatorship such as that of the USSR would be much nearer slavery than it is, and the liberation of women would be unlikely to have happened where it has. Slavery in the South would ultimately have withered when the electric motor and the car arrived, only a generation after emancipation in Russia and the USA.

31   The borderline between 'conscious' and 'unconscious' slave-breeding was in the provision of good stock-getting sires for *all* female slaves (except relations) on a plantation. This might result in 25–40 per cent of all females producing a baby each year. Left to themselves, with 'marriage' and monogamy, only about 10–15 per cent of the female slaves would be capable of a live birth every year.

32   One indicator of the contemporary state of hygiene in the United States, both North and South, was the loss from disease in the Civil War. In the Northern armies more than twice as many died of disease as did in *all* battles. 'Ignorant city trash,' said the Southerners. In the South, 60,000 died of disease, 94,000 of *all* battle causes. Most of the 'disease' was avoidable enteric disorder of one sort or another. The Southern Army resisted these problems better than the boys in blue because the Confederates were usually in a permanent state of dysentery, mostly caused by inadequately cooked food. Thomas Keneally's *Confederates*, Collins, 1979, gives a vivid picture of this problem.

33 Disaffected slaves had ample opportunity to extract revenge. It was almost impossible to discover who had added powdered glass or excrement to food, or put itching powder in beds, or minute slits in 'waterproof' boots. Slave-owners spent a lot of effort on keeping their household slaves sweet, and the irony was apparently lost on them that they, the owners, were as much in bondage as were their slaves.

34 Some states were extraordinarily short-sighted about good and bad money. Traditionally, settled areas like hard money, while new, frontier sections prefer easy, soft conditions. Everyone knows what happens when governments are permissive about money: values are destroyed in a torrent of inflation. By then, the beneficiaries have moved on, or, in the nineteenth century, converted to gold. See J.K. Galbraith, *Money: Whence It Came, Where It Went*, Deutsch, 1975, which is very good on the antebellum position.

35 Since everyone on the frontier could make his own whisky, and get drunk for a few cents, it became the habit in the South and West to vote for 'local option'. this meant that County A would be dry, but next door in County B beer only might be legal, and further on, in County C, everything could be purchased. At one time neighbouring towns would be on opposite sides of the temperance fence – between East St Louis and St Louis, for instance, traffic across the Mississippi was intense, especially on Saturday nights.

36 Each 500 lb bale of cotton left behind twice that weight as residue. Of the 1000 lb, 450 lb consisted of fibrous material, of which some was bran, edible by cattle. But the whole of this fibrous material could be used as fuel without destroying any of its value as a fertilizer, for which purpose the ash could then be used.

Of the remaining 500 lb, 150 lb would produce: cottonseed oil, both edible and inedible, cooking fat, soap, miner's lamp oil, and even some salad oil; 400 lb would be cottonseed meal, or 'cotton cake', for use as an animal feedstuff or fertilizer. It contained 45 per cent crude protein, or nearly 7 per cent nitrogen, a very valuable balancer for grain in cattle-fattening, or a fine, easily used organic fertilizer.

The trade in these residues did not develop until the 1870s, after which in some years the by-products were worth more than the cotton itself.

37 In largely cashless societies, which have always existed in some place at some time, the payment of rent in money has proved impossible. In Europe, the metayage system gave the landlord about a quarter of the crop as his 'rent'. In the South, after the Civil War, the whole organization of society had broken down, as in Occupied Europe in 1945. Above all, the currency of the defeated side was worthless. Faced with an almost cashless and creditless condition, the landlords of the defeated South 'let' their land for a share of the crop, sometimes to negroes, but in the conditions of the 1870s mostly to poor whites. As it turned out, revival was more rapid than could be imagined in 1865.

38 The intrusion of the moral issue made compensation very difficult to accomplish. In 1833, in Britain, the abolitionists argued fiercely that the wicked West Indian slave-owners should in no way profit from their turpitude, and the same sort of people used the same sort of argument in the United States. In the end, the North spent nearly 12 billion gold dollars on the war, or six times the highest cost ever postulated for compensation.

39 The iniquity of slavery was greatly increased, said the abolitionists, by the practices necessary to keep the slaves in servitude. Not all of these practices – prevention of literacy, the curfew, whipping, cruel and unnatural death (burning alive, for instance, for raping a white woman) – applied in every state, but the pass laws and the fugitive slave laws were universal. A slave off his owner's property obviously had to carry a pass, or he was presumed to have 'escaped'. Under the fugitive slave laws, escaped slaves could be pursued into a 'free' state, and this collusion was much hated and resented by Northerners. It was probably the treatment of the fugitive slaves which led to a kind of cold war between North and South in the thirty years before the shooting war started. After 1853, the slaves were aiming for Canada, where they were safe and free: there was said to be an 'underground railway' of sympathisers who helped slaves reach Canada. Very few did. Distance militated against success – it is nearly 700 miles from Atlanta, Georgia to the nearest point of Canada. See also note 49.

40 The people of the Appalachians and Alleghenies in Virginia and the Carolinas were extremely reluctant heroes in the Southern cause, and quite heavy resources of manpower were used to conscript the able-bodied males. The men concerned were often the only workers on their tiny holdings and the only means of support for several women and children. Possibly as many as 50,000 men deserted. In Virginia west of the Shenandoah mountains the hillbilly population was in a majority and successfully seceded from the Confederacy, becoming in 1863 the independent state of West Virginia. The proportions of support for the North amongst the people of West Virginia can be gauged by the numbers in each army: Union, 36,000; Confederate, less than 10,000. After the war, West Virginia boomed as a primary producer of coal, oil and natural gas. Today, it is being restored with federal and state money.

41 The Northern workers were encouraged to despise and fear the slave as a rival and as a factor in keeping wages down. In fact, of course, in no way did the slaves in general compete with Northern workers, and the ceiling on wages amongst the unskilled was the result of immigration from Europe, and not of competition from Southern slaves. Yet few immigrants went to the South, because they disliked the idea of a slave society.

42 In Egypt the great group of buildings in Cairo called the Citadel was largely built by Crusader prisoners of war, from half a dozen European countries, between 1175 and 1200. With the aid of slave workers their captor, Saladin, erected the most enduring monument of his time – much more enduring, for example, than the Bridge on the River Kwai. The PoWs were neither castrated nor returned home, so their blood must flow in the veins of the people of Egypt today.

43 After about 1650, European agents deliberately encouraged African chiefs to provoke warfare on their neighbours. This hypocritical exercise made it easier for everyone to accept that the resultant captives were prisoners of war, and that it was therefore morally permissible to enslave them. The losses in dead and wounded were, however, sometimes horrifying, even for those who could not count.

44 As a result of the Civil War 600,000 men were killed or died prematurely. This was 2 per cent of the 30 million whites on both sides, a higher proportion of fatalities than the British suffered in World War I, and ten times greater than British losses in World War II. These massive losses mask disturbance and disruption in a much larger proportion of the population. The experience, however measured, was traumatic. It was probably the making of the United States, not only because Secession had to be defeated for the Union to be a reality, but also because, without the waste, disease and death, nationhood would have as much real meaning in the

United States as it has in Argentina or Brazil – which is not very much. Whether or not national consciousness is a good or bad feature of life is not an issue. In the USA the sense of nationhood was reinforced and made permanent by the War between the States, by the bitter-sweet memories, by the sense of loss, by the tragedy which older nations have in abundance to form their sense of communal inheritance. It is the tragedy of life which forms a nation, not the trivia.

45   Greycloth is undyed, unfinished cloth straight from the weavers. It is ready for bleaching, printing or whatever, and is a standard commodity, like 'lint' (cotton wool) or 'yarn' (thread).

46   There were apparently no more than fourteen readers of the *Economist* in the USA in 1860–1; none of them lived south of Richmond. Today, there are said to be 200,000 readers, all over the place.

47   Yarn production in the United Kingdom fell from nearly 500,000 tons in 1860 to 200,000 tons in 1863, the worst year. As the UK then produced more than the rest of the world combined, the effect of the blockade on American cotton exports can be said to be worldwide.

Most of the workers in Lancashire were laid off, or on short time, and very hungry. So they were in New England and the middle Atlantic states, but the Civil War was an American fight, not a British one.

48   The most obvious candidate for being today's top 'Confederacy' is South Africa, not because the blacks there are treated as slaves, but because the white South African ethos has little future. But a 'Confederacy' exists wherever the future is feared, the present is based on false premises and the past is viewed with nostalgia. Both Greece and Turkey are 'Confederacies', each inducing the characteristics in the other. Certain British cities are 'Confederacies', for the same reasons as South Africa. A great many once-great universities, corporations and other human institutions have also become 'Confederacies'.

49   The 'Peculiar Institution' of slavery was a cant Southern phrase first recorded in the *South Carolina Gazette* in 1852, but used in conversation before that date.

50   The first really serious black/white race riots in the North were in Detroit in 1943, not in the 1960s as is often thought. On the whole, cities such as Atlanta have been very much more successful in their race relationships than have cities of similar size in the North. As soon as an appreciable black middle class is established, the divisions in society can be expressed in the usual terms of income, occupation or educational attainment rather than race. If the black is at the bottom of the heap, his position as an unskilled low wage-earner can always be explained by the fact of his blackness.

The Potato

The Potato
and Irish America

So far this book has looked at the social and economic effects of various plant transfers primarily in non-European countries, in warm climates and far away from everyday use in an advanced country. This chapter, however, deals with the far-reaching effects of the potato upon the history of two European countries and a third mostly peopled by Europeans.

The potato comes from the high Andes, 8000 feet above the maize or Indian corn line. Like maize, the potato was the staple starchy food of the Incas. Four hundred years before the Spaniards arrived in the early sixteenth century, the Incas were a small tribe settled around Lake Titicaca, the world's highest lake. By the time the conquistadores challenged their civilization, the Incas ruled an immense strip of Andean territory from Quito in modern Ecuador in the north, to an area south of both modern Santiago and Valparaiso in Chile, a stretch of over 2000 miles. Their epicentre was Cuzco, one of the great silver mines of the world. The attraction of their economy for the Spanish conquerors was gold and silver, but the Incas' real gifts to the world were maize and the potato.

Various tales are told about how the potato was transferred to Europe and North America. A slightly sententious Victorian story was told by the historian Pink in 1879. He suggested that Sir Francis Drake, on his voyage to pick up the surviving settlers from the failed settlement at Roanoke in Virginia in 1586, gave some potatoes obtained in the West Indies to Harriot, Sir Walter Raleigh's agent. Harriot planted them at Youghal in southern Ireland in the next year, and offered Raleigh the product to eat in 1590. Unfortunately, Harriot gave him potato seed rather than tubers and Raleigh, not unnaturally, experienced indigestion rather than nourishment.[1] Despite this inauspicious start, potatoes had become a staple of the Irish by 1625. Unfortunately everything about this story except the last sentence is untrue.

Drake certainly knew about the potato: he had been offered it in Chile in

1577 during his voyage around the world. Potatoes were not indigenous to either the West Indies or Virginia, and originally grew only in the Andean highlands. Herbalists such as Clusius, Banlin and Culpeper all linked the potato to Europe at a date earlier than 1580. Potatoes floating ashore from the wrecked Armada in 1588 were alleged to have colonized western Ireland. They were certainly in use in Spain, Italy and France before 1600, but there was much early confusion between the true potato of the genus *Solanum*, the sweet potato, *Ipomoea*, and the yam, a semi-tropical plant which will grow in the Mediterranean and is a member of the Dioscoraceae family. What is certain is that the white potato was a mere garden crop or cattle food in every other European country before 1650; in Ireland, as already mentioned, it had become a staple.[2]

What we now know to be most likely is that in 1586 Drake aimed to intercept and capture the annual Spanish treasure fleet in Cartagena on the Caribbean shore of modern Colombia; he missed it by twelve hours. Carrying on to Virginia and England, he carried with him not gold and silver looted from the Spanish, but a few potatoes, mere curiosities, which had been obtained as ship's stores somewhere in the Caribbean. These potatoes, says Pink, were of far greater worth than all the gold and silver in the treasure fleet. The immense consequences of the way in which they were subsequently used or abused in Ireland, however, involved not only that country but also Britain and America; and most of those consequences history could usefully have done without.

Ireland before the potato was conditioned by three factors, the most important being its geographical position. It is an offshore island, on the edge of another island on the edge of the continental shelf of Europe. At the nearest points Scotland is only 14 miles away and Wales less than 50, while England is more than 100. Ireland is too far from and yet too near to continental Europe.

Placed several hundred miles to the west, Ireland would have been left alone except by a body of men so strong, so determined, so steadfast that they would have brought to Ireland a permanent and ordered body politic, as in Iceland. Positioned nearer to England and France by being a few hundred miles further southwest, Ireland would have become part of the Roman Empire. Her nearest neighbours were in the islands of Scotland, and traffic across the intervening sea must have been possible, though not common because the Scots and Irish were too poor to indulge in much trade. The dozens of islands in the western archipelago of Scotland, with which both Ireland and Scotland proper have had intermittent contact through the ages, contained no wealth, nor vigorous populations, nor natural resources. The Western Isles have always been noted for their unproductive soil, savage weather and magnificent scenery, and for the dominant sea which rules the lives of all living creatures, including fish, who are by far the most numerous inhabitants of the continental shelf and the only true natural resource.

The ancient Alexandrine mathematician, astronomer and geographer, Ptolemy, whose second-century view of the world was to remain unchallenged until the Renaissance, was uncharacteristically sketchy about Gaul's neighbours. Scotland (Caledonia) was placed to Britannia's northeast, ending in Thule (the Shetland Islands), which he located somewhere near modern Bergen in Norway. Of the position of Scandinavia he was ill-informed, and of the Baltic ignorant. Ireland (Ivernia) was placed further north than any part of Wales, its shape wrongly drawn. Unknown to the Roman conquerors of what is now England, Ireland lay somewhere offshore from Anglesey, whose Druids may have been in touch with their fellow Celts. But the Romans, a Mediterranean people to whom seafaring out of sight of land did not appeal, left Ireland to the mists and fogs of ignorance.[3]

Ireland's second conditioning factor was climate – a much greater problem for an ancient civilization than for a modern one. There is a school of thought which claims that the summer climate in the Eastern Atlantic became cooler and cloudier about 500 BC. Previous to that date Ireland may have been a bread economy, it is said; but for this belief there is no evidence whatsoever either way.[4] What is quite clear from modern observation and agricultural experience is that only in the southeast of Ireland was it possible to grow the kind of cereals from which bread can be made – wheat or rye. Though a rather nasty unleavened bread can be made from barley, all cereals can be made into biscuits or cakes of a sort, such as oatcake, and all can be turned into porridge or gruel; but it is only since improvements in plant breeding, mechanical invention and husbandry were made that any farmer could contemplate growing bread-grain in most of Ireland.

There were other handicaps. Iron ore is not found generally in Ireland, so that the Iron Age may have existed only in what is now Ulster. There was no copper or tin in prehistoric times – though there was a little lead ore containing silver, and gold in the Wicklow Hills. Ploughs, even the wearing parts, would therefore have to be made of wood. This would mean cutting a new ploughshare point every hour or so, which effectively restricted cultivation to the malleable soils, and then only during the easiest part of the year.

To the lack of iron, which left Ireland virtually a Stone Age economy until AD 1000, must be added another handicap. Within living memory, horses drew ploughs 'by the tail'; this practice, employing one rope instead of a set of leather or rope harness, is not only very cruel but also inefficient. Draught harness – comprising a breast strap and collar – was known in China by the sixth or seventh century. It did not reach western Europe until *c*. AD 1000, and was not seen in most of Ireland for another six hundred years. Harness was one of the technological benefits brought to most of Ireland only by the Scots and English after that time, and it is quite extraordinary that this inefficiency continued long after every invader had introduced or reintroduced harness. Tudor Englishmen, not noted for their compassion, complained about the practice of draught by the tail, yet some descendants of each generation of settlers could be found, a few years later, succumbing to the local custom. Not for the first or last time the Irish could be accused of a particular practice, yet history finds that the son or grandson of the critic was indulging in the same practice at some future date. The capacity for the Irish to absorb the foreigner, and for the foreigner to go native, can never be over-estimated.

Whether or not the Irish grew cereals widely, they seem not to have solved the problem of threshing corn. To the astonishment and dismay of every observer, even into the nineteenth century the Irish would not cut their corn with a sickle or scythe and then allow it to ripen in the stook and be carried away and threshed. Instead, the ripening corn was cut or pulled up and then burnt, sometimes roots and all. This unique method of harvesting was long cited as a supremely Irish form of behaviour. (The eighteenth-century English were very rude about Irish thought processes, and indeed the 'Irish' joke still flourishes today.) But this practice, which incited ridicule and contempt from foreign visitors for nearly a thousand years, was not, perhaps, as odd as appeared at first sight. In an uncertain climate, corn a month away from 'normal' harvest in, say, the east of England would take much more than a month to ripen in damp Ireland. Corn a fortnight from harvest might be as far as nature would ever allow, and this stage might not be reached until November. To scorch the corn might be the only way to render it edible, by man or beast, neither of whom would benefit from unripe corn. The modern farmer who, in a wet summer, cuts his corn immature and then dries it artificially is employing the same basic husbandry.

The third and greatest of Ireland's problems was the absence of a personal relationship between men and the soil that they worked. The difficulties of cultivating the soil due to a lack of iron, the unsuitability of the weather for grain-growing without modern techniques and machinery, and, worst of all, the fact that serfs were attached, not to the soil, but to the lord, marked Ireland in AD 1000 as unique in western Europe – the only country with all three deficiencies.

Slavery had become feudalism over most of Europe by the ninth century AD. The king held in feu the land of the great earls; the great earls held in feu the land of the lords; the lords held in feu the manor, which had attached to it so many serfs, who worked the lord's land for so many days. The important factor is that the whole edifice was based on landholding, from the humblest serf (attached to the land) to the greatest earl, who held a county from the king and owed so many days' service a year in exchange, and was in turn owed so many days' service in lieu of rent from his feudal vassals. But feudalism did not derive only from Rome: the Saxons, Jutes and Angles all adopted the same system even though they were, like Ireland, outside the Roman Empire; not so the Celts.

In the pure Irish culture of the period between the Christian mission of St Patrick in the fourth century AD and the destruction of that culture by the Vikings in the eighth century, a period of which we know little, the law was that of the Celts and known as Brehon. The great men were called 'kings', of whom there were 150 or more in a population of about 500,000 in AD 1100. They usually did not admit the existence of a super-king, but such figures arose from time to time as a result of some foreign threat or other. Kingship was usually, but not always, hereditary. The children of the king were not educated at home, but sent to board with another king. This exchange of children was supplemented by the exchange of other hostages who appeared in every king's hall. The king also had attendant wise men, poets, jugglers and personal bodyguards, the last being slaves whose lives had been spared in exchange for exaggerated personal loyalty. The whole system was based on personal relationships, not that of the relationship between man and property, man and land.

The diet of the Irish at this time included oaten cakes, cheese, curds, milk, butter, cattle blood,[5] eggs, and in prosperous areas pork, beef and lamb. Game abounded, since there were 40 acres per head of bog, forest and cultivated land, as well as unpolluted rivers full of fish and bees everywhere to produce the honey which was almost a sacred food for the Celts. Mead and barley beer were drunk by the better-off; wheaten bread was a great status symbol and a rarity for everyone except the kings, since the wheat was either imported or grown only in the southeast of the island.

There were no towns – the Vikings were the first to build any – no roads and no coinage. The unit of value was a cow, and the cow unit was known as a set; gold, silver, bronze, tin, clothes, pigs, horses and slaves were all valued in sets. Slavery was only abolished in 1171: a female concubine, unless pretty, was worth three sets, a king's daughter twenty, and marriage by purchase was universal. Polygamy continued until about 1400. Bastardy was considered no bar to succession, and there was no idea of illegitimacy.

This cattle-value economy could have existed in peace for hundreds of years at a level at which everyone had enough to eat and enough land to support themselves. There was a surplus available for export, to exchange for metal, cloth and precious jewels, and for the support of the noble but unproductive professions of priest, poet and law-giver. The evidence of the surviving gold ornaments of the period, such as rings, bracelets, armlets, torques, crescents, necklaces and coronets, points to a society with an aptitude for decorating its richer and more beautiful women, and with the necessary wealth to do so.

The surplus also enabled endless civil wars to be waged – the summer occupation of the fighting men, who went on campaign when the cattle went up to the rough grazings on the mountainside. If Ireland had never been invaded by any foreigner, Viking or English, it is difficult to see how the Golden Celtic Age could have produced an organized, unified society like that of Norman England. The Celtic predilection for petty warfare was too strong, and the inability to develop the institution of monarchy in place of the Gaelic chiefdoms too marked. This theory of Brehonic inadequacy can of course never be proved; nevertheless the attachment by blood and sentiment to human relationships rather than to land made Ireland almost unique not only in the Middle Ages but also for several centuries afterwards.

The Irish and some of the West Highlanders sometimes claim that the clan system of the Scottish Highlands was similar to that of Ireland. Though there may have been similarities, there are also considerable differences: the Scots were probably much more territorial than the Irish; the inhabitants of Scotland had traded with the continent since the beginning of recorded history; Scotland was affected by the Roman occupation of Britain, and, for a period, of southern Scotland too; and before 1200 the Anglo-Normans established in southern Scotland the shire system, with all the implications of feudalism. The Scots were much less attracted to the personal connection between vassal and lord than they were to the trinity of land, lord and people.

The same exaggerated loyalty of person to person, of king to numbers of slaves (not serfs), and of slaves to kings, and its attendant interminable warfare, led also to the premium placed upon a high birth rate. It was encouraged in the feudal societies of the rest of Europe, because one measure of wealth was the number of dependent retainers. Yet feudalism contained its own in-built safeguards on population: the attachment of the serf (not slave) to the land meant that however many mouths he was responsible for had to be fed from a fixed area. No such automatic limitation existed in Ireland: kings could always hope to conquer more land, enslave more prisoners, steal more cattle. Instability was ensured by the absence of any true feudal system.

All over feudal Europe during most of the Middle Ages there was a natural connection between the land, the people living on it, and its capacity to support them. This idea was as accepted and unspoken as the natural assumption that the land would support so many beasts of the field. The importance and stabilizing effect of this universally accepted ratio gave Europe an equilibrium for hundreds of years. This situation was broken from time to time by several factors, of which the Black Death of 1340–60 was the most dramatic, the New Learning the most intellectually demanding, and the rise of trade, industry, towns and commerce the most insidious.

Of all the rural areas of Christian Europe, Ireland alone did not share this equilibrium. No census of any value was taken in Ireland until 1821, but the available estimates going back to the sixteenth century confirm the ability of the Irish to increase their population far beyond the resources of the agricultural conditions of the times. In each of the three centuries before statistics were prepared, the sixteenth, seventeenth and eighteenth, and again in the nineteenth century, when the figures can be proved, the population trebled.

Social conditions were dominated by the Celtic ethos; so was the Church. The Irish Church was the child of Brehonic conditions in Ireland, and as much a victim (or beneficiary) of the lack of settled organization as any civil institution. Before St Patrick's arrival in about 432 the Druids, who maintained in Ireland as in pre-Roman Gaul or Britain the role of wise men, magicians and witchdoctors, refused to learn to read and write, and dismissed European written history and evidence as unnecessary and confusing. Druidic rites were absorbed by the Irish Celtic Church, as were the Roman festivals by the Latin Church, the Levantine practices by the Greek and so forth. The Irish Church, non-territorial, non-episcopate, non-hierarchical, grew up very much on its own, a unique Gaelic institution, an inspiration, and a source of profound annoyance to every Pope.

The Gaelic Church was founded upon monasteries, holy places and hermitages – not on parish churches as in the rest of Europe. No dioceses to speak of existed until they were imposed by the Anglo-Normans. But there was a plethora of 'bishops': over a thousand in the sixth century, for example, in what is now County Louth. Each king might have several bishops at his court. In some monasteries every monk was a bishop and some were saints before they died.

Faced with a surplus of men, often quarrelling, the Irish Church encouraged missionary activity. Monks or bishops from Ireland went to Iona, most of Scotland, Lindisfarne and the north of England, southwest France, Italy and Germany, as far north as Iceland, as far east as Hungary and Poland. Whenever there was a strong Roman organization, the Irish missionaries were discouraged. But they established the independence of each congregation and the peculiar flavour of the body of Celtic belief which wove together the Roman legacy with the early Christian attitude to the Church, sprinkled liberally with the Druidical mix of magic and mystery. Wherever they went in post-Roman Europe, the Irish brought with them a Celtic grandeur of the spirit. But they were not popular. Services were held in Gaelic, and continued to be so until the fifteenth century. Bishops were married. They were accused of converting slaves without the consent of their lords, of consecrating as bishops unordained laymen who had only just been converted, of permitting bishops to be consecrated by another, autonomous, bishop instead of by general agreement – no wonder there were so many. They learnt to read and write, and to engage in learned arguments about the date of Easter. This massive outburst of activity took place during the four centuries between the arrival of St Patrick and the establishment of the Norse government in the area round Dublin in 848. During the next two centuries the Norwegians, the Danes, the native-born Irish, and Irish- and Scots-born Vikings fought over the carcass of Ireland and southwest Scotland, and robbed and destroyed the Church, reducing Ireland in particular to helpless poverty.[6]

The invaders of Ireland, whoever they might at any time be, showed particular animosity towards the Christian religion. The sacred monastery of Armagh, centre of faith in all Ireland, was sacked and totally destroyed ten times in two hundred years, and each time totally rebuilt. Once it was attacked three times in a month, though the sufferers were not then the inanimate stones, but the monks trying to rebuild the place. To escape from continuous attacks, Irish monks and scholars fled in large numbers to the continent; with them they took their books as well as their learning.

Just before the year 1000 there arose a great Irish leader, Brian Boru, who became the paramount king in about 1002. He encouraged the administration of justice and the building of roads, bridges and castles, and was especially kind to bards and historians, who have consequently left us a good account of this civilized man. In 1014 Brian and the Irish fought a great battle at Clontarf against Norsemen from Ireland itself, from Scotland, from the Isle of Man, and from as far afield as Orkney. The Norsemen were defeated and the possibility of a Norse hegemony in Ireland was forever destroyed, but Brian and two of his sons were killed and the Irish relapsed into the continuous anarchy of civil war.[7] It was in this state that the Anglo-Normans first contemplated their neighbour.

If the Norsemen had not been strong enough to subdue Ireland, certainly the Anglo-Normans were not, and an unhealthy situation kept repeating itself over five or six centuries. The native Irish, plus the half-Irish and the quarter-Irish, were never politically united enough to live in peace. The anarchy of hundreds of warring chieftains dismayed every

English king with strength and time enough to attempt any solution. But, faced with a foreign invader, the Irish united for long enough to defeat him, and then relapsed into the luxury of internecine, nearly always unrecorded, civil war, rape, slaughter, pillage and destruction, in which little survived that had previously stood or lived or grown in the lands of the defeated. The Pale was the exception. This area around Dublin was dominated by the English, as it had been by the Norsemen; it had the best arable land in Ireland, and was known to the natives as 'The Land of Promise'.[8]

Contrary to the belief of the Celtic revivalists of the nineteenth century, the Anglo-Normans were drawn to Ireland not by hope of gain, but by the realities of geography, and the impossible task of dealing peacefully and diplomatically with the Irish. Nor can the Anglo-Normans be credited with the destruction of Celtic Ireland. Long before them the Norsemen – the Danes or Swedes or Norwegians, together with their allies and relations, the Scots-Norse and the Irish-Norse – had brought the Golden Age to an end. No English leader before Cromwell had sufficient motivation or force to affect Irish political or social conditions except in the Pale around Dublin. Post-Norse Irish conditions, left alone, would have made a settled society very difficult in any case. The Brehonic laws made Celtic Ireland a country without the means of national development, and without the capacity or inclination to set up the kind of larger organization needed for the future.

If the Irish had united in a hegemonous country, if there had been a feudal system founded on land, instead of tribalism based on personal loyalty, if sheep and cattle had not been favoured over grain, if the Irish Church had not been so disorganized as to attract the blessing of every Pope upon the Norman or English invader – if all these situations, and others, had not existed, then the history of Ireland would not have been as it was, and the right conditions for the adoption of the potato would not have existed.

In the five hundred years between Henry II, the strongest European king of his day, and Oliver Cromwell, the strongest ruler in a western Europe exhausted by the Thirty Years' War, none of the considerable English effort expended in Ireland achieved any permanent result. The English kings, sometimes aided and abetted by the Papacy, sought to impose the shire system of civil organization and the parallel diocesan bishops, while the Irish resisted both except in an area in the southeast of the island. The ancient nobility of Ireland objected to tithes, Peter's pence, canonical decrees and the observance of non-Irish festivals.[9] This Celtic resistance was to continue after Henry VIII had suppressed most of the monasteries and had given their lands to Irish, Anglo-Irish and Norse-Irish nobles, and the friars had taken over from the dispossessed monks and nuns. The monasteries proved resilient: suppressed where possible by Henry VIII; revived under Mary; ignored by Elizabeth; suppressed again by Cromwell; permitted to be revived under Charles II.

Language was another source of strength to the Gaelic natives. Except in the Pale of Dublin, most religious services were conducted in Gaelic; it was the English who translated the Mass into Latin, and that, curiously enough, happened as late as the fifteenth century. Even in the time of Henry VIII, perhaps half the Irish took Mass in Gaelic, and the Celtic Church, which had been fiercely nationalist and independent before the Reformation, remained a focus for resistance both before and after the English became Protestants.

After Elizabeth's death Ireland, relieved of an aggressive English sovereign, recessed further into factionalism. By about 1620, by which time the potato had begun to be very important, all the elements necessary for its adoption were present: a weak executive, an absence of feudalism, virtually no grain cultivation outside the Pale, a cattle-based economy, little respect for land ownership, land rights or territorial divisions, and almost continuous civil disturbances ranging from brawls and riot through murder and arson to full-blooded revolt.

Of all these aids to instability, the one most inclined to cause misery was the absence of any tradition of land tenure. If a native landlord class is not established, it follows that there will also be no tradition of true tenantry, with all the security that true tenant rights confer. When an alien landlord class establishes itself in a conquered country, it may be inconvenient to establish the natives as tenants of the new landlords. Alien conquerors of the more modest rank may become the new tenants. If a landlord–tenant system already exists, the conquered country is much more harmoniously occupied by the new rulers. In time, the conquered class absorbs the conquerors. This had happened in England after the Norman Conquest in 1066. It never happened in Ireland. The conquerors seized the land and ignored the occupiers.

The 'natives' were pushed into the least favourable areas and the least favourable fields, up the mountains and into the bogs. At some point they reached the limit of all cultivation except that of the potato.

The first great side-effect of the English Reformation in Ireland is not generally recognized: it allowed both the English and the Irish to blame religious change for what would have existed anyway. The Irish became Latinist in religion while the English became Anglican, but even if they had both remained within the Roman Catholic Church, England would have remained Latinist and the Irish Celtic, and the basic quarrel, which

has never had very much to do with God, religion or Rome, would have continued.

The other point which is usually ignored is the small percentage of the population which spoke English. As in the Highlands of Scotland, Wales, Brittany and Cornwall, the local Celtic language was the spoken word of nearly all natives, nobles included. The educated might know Latin better than English, and until quite a late date the written word would have been in Latin. It is significant that the Irish of all classes regarded the Anglo-Normans as Saxons and spoke and wrote of the English by that name as late as 1600. In truth, they were right. The real quarrel was not between Catholic and Protestant, peasant and occupying power, old nobility and new men within the Pale. The real quarrel was, and is, archaic, primeval, between Celt and Saxon.[10]

The 'Scots' offered hope in 1603. Some of the Irish recognized that they had once been 'Scoti', while the modern Scots had been Picts.[11] The new English king, James I – James VI of Scotland – had another advantage in Irish eyes. He was not Elizabeth, a name the most ignorant peasant had learnt to hate.[12] But his love of order was not tempered by the wisdom enjoyed by his great predecessor; on the other hand, his early initiatives in Ireland were blessed with luck and apparent success. He 'planted' English and Scots in Ulster, following on the less well-known plantation of Munster by Elizabeth.[13] He encouraged the Corporation of London to anglicize and fortify Derry. He made a large number of Englishmen 'baronets' if they subscribed to the plantation. He increased the Irish House of Commons from 122 members to 232, and guaranteed an in-built, permanent Protestant majority. He made much noise and commotion about banishing Roman priests, but they returned; about making the native Irish elite attend Anglican churches, but his extra-parliamentary decree proved to be illegal; and about the introduction of English land tenure and the English shire system, which succeeded. This last drove thousands of natives out of the normal economic system, the cash economy, and they became dependent on the potato. The 'Flight of the Earls', which had preceded many of these measures, left much of the northern native population leaderless and looking to Spain or France for salvation.[14]

In the generation before the execution of Charles I in 1649, the situation in Ireland was even more chaotic than during Elizabeth's Irish wars. Even Thomas Carlyle, writing in the nineteenth century, was puzzled: 'There are,' he said in the historic present,

Catholics of the Pale, demanding freedom of religion, under my lord this and my lord that. There are Old-Irish Catholics, under papal nuncios; under Abba O'Teague of the excommunications; and Owen Roe O'Neill, demanding not religious freedom only, but what we now call repeal of the Union and unable to agree with Catholics of the English Pale. Then there are Ormonde Royalists of the Episcopalian and mixed creeds, strong for the King, without covenant; Ulster and other Presbyterians, strong for the King *with* Covenant; lastly, Michael Jones and the Commonwealth of England, who want neither King nor Covenant.

The complicated and confused puzzle could only be solved by bloodshed. This was provided by Cromwell, normally a merciful man, but not in Ireland. Like other Parliamentarians, Cromwell was angered by Charles I's use of Irish soldiers in the English Civil War, and made sure that the Irish would regret their intervention. He set out to destroy the remaining tribal organization of the Irish, as did the English in the Highlands of Scotland a century later after the 1745 Jacobite Rebellion. The Irish were driven to 'hell or Connaught', the least fertile of the four provinces, in those days consisting mostly of bog or rock. The plantation Scots in Ulster were reinforced, as was the English landlord class in Munster and Leinster. The Cromwellians had every intention of making the Irish starve.

The potato, *Solanum tuberosum,* is a plant native to high altitudes, with thin soils and short days, low temperatures at night and a dry atmosphere. In western Europe it flourishes in moist, cool atmospheres, with relatively long days and warm nights, and enjoys a deep, friable soil. In poor soil, potatoes are much more suitable than grain crops. Few tools are needed; in an emergency, a crop can be grown and harvested with one's bare hands. Nor do potatoes need threshing, grinding or baking; a pot and a peat fire are enough.[15]

The gene bank of the potato contains formidable variation, and by natural selection, hybridization (either accidental or man-induced) and survival of the fittest in the new circumstances the potato can go through a number of permutations which effectively amount to a continuous process of adaptation. The peasants who survived most successfully on the monoculture established in Ireland were probably those who by observation and trial and error had gathered stocks of potatoes which were best adapted to the conditions found in the poorer areas of the island. The Irish people needed every adaption of which they – and the potato – were capable.

We do not know how many Irish were killed in the twenty years between 1640 and 1660. They were killed directly and deliberately by Cromwell at Drogheda; they were killed in leaky ships, which drowned many of the 'criminals' transported to America or the West Indies;[16] they were killed by starvation at home or by tropical disease abroad; and they were killed by the destruction of the livestock which was their only wealth. The population was perhaps halved, and the half that remained had no tools, no livestock and often no land. One fact is known: without the potato, none would have survived. For, pushed into the least fertile province with no cash and no means of earning any, and with no chance of growing bread grains on those inhospitable bogs and mountainsides, there was nothing but the potato to keep the Irish nation alive, and no way of growing the potato except in the lazybed. Ironically, of all the havoc wrought by Cromwell in Ireland the by-product, the lazybed, was in the end the most damaging. Cromwell was an obvious enemy; but the lazybed was a friend which ultimately resulted in the deaths of more Irishmen than he ever slaughtered.

The lazybed can be prepared on any type of ground, almost anywhere. The land need not be flat, and stones do not inhibit potato-growing as they do in normal fields. As the bed is self-draining, a bog is almost as suitable as a mountainside. A strip of land, from 2 feet wide in very wet soils to as much as 6–7 feet wide in dry soils, is spread with whatever manure may be available: seaweed, the rotted turf of an old house, or dry peat. A ditch is then dug on either side, and the earth from it is thrown up on top of the manure. The bed is now self-draining and the potatoes have been 'ridged up' before they have been planted. The tubers can then be planted with a dibbler. Alternatively, the setts may be placed on the manure and covered with earth as the trenches are dug.[17]

A strip between 500 and 800 yards long would provide enough potatoes for one family. Supplemented with milk, pork, bacon, cheese or the blood of a cow, they would provide a balanced diet for a typical family. When times were bad, there would be no supplementary food, the peasant and his family would need a larger area, and the bed had to be lengthened.

The lazybed had many advantages. Half an acre could provide in a 'normal' year for a 'normal' family. Even though it was unfenced, it was secure against wandering stock, and unattractive to marauders, whether soldiery, neighbours or the hostile people of another tribe or clan; it was immune to frost, well-drained and well-manured. After the potatoes had stopped growing the lazybed came into its own, since it could be used as a clamp for storing the tubers, which needed only to be dug as required and then go straight into the pot. In other systems potatoes must be lifted and stored in a frost-resistant clamp or building, which involves more work.

If the worst came to the worst and the peasant, having spent the summer in the hills, was unable to return home at harvest time, the lazybed would be full of potatoes in the following spring. He could thus eat well at a time of normal dearth and still leave tubers in the ground for next year's crop.

The lazybed was so-called because not all the ground had to be tilled: about half the area had inverted sods placed upon the unbroken soil. Nevertheless the English, through ignorance or malice, thought of the lazybed as an idler's way to grow potatoes; it became a drawing-room joke about the Irish, with obvious variations of an elegant, if raffish, nature.

No sleek, well-fed, censorious Englishman knew that the Amerindians in modern Peru had adopted the system centuries before. Nor do most Englishmen know that the lazybed is still employed in the Andean uplands, 8000 miles away, as an appropriate method of culture. No one knows whether the method came to Ireland with the potato, or was evolved by the Irish in the face of Cromwell's actions.

Although the lazybed met a need, it also produced some wholly unfavourable results. If the potatoes produced in a lazybed were a family's only source of food, the cultivation of those potatoes and the cutting of the turf needed to boil them and to augment the heat in the cabin produced by the bodies of the man, his wife, children and a cow, plus perhaps a pig or two, would take at most ten to fifteen weeks in the year. The only cash a man would need would be to pay his rent. This could be earned at harvest time, helping the (often absentee) landlord to garner his cereals. These cash crops grew on the better, more easily cultivated land, and the grain was often exported. Since the wage was only needed for rent it could in fact be covered by setting off so many hours' work against so many square yards occupied by cabin, lazybed and grazing for an animal, if any. This was called 'truck' and led to a shortage of cash. In the eighteenth and nineteenth centuries the amount of coinage in circulation in Ireland was only 20 per cent, per head of population, of that in Egland.

Sometimes there was no work at harvest time, so begging became endemic, though from whom you may beg in an impoverished land without any cash economy, no one has explained. Since, however, all reports of widespread begging came from the literate, usually a visitor or a member of the English Ascendancy,[18] and since presumably they had money, the beggars must have selected their victims with discrimination and commonsense.

The great tragedy of Ireland, which reached its climax in 1846, would probably have arrived without any further help from the English once the lazybed had been widely adopted. A couple wanting to marry had only to throw up a cabin, which took them or their families less than a day. A turf cabin was the traditional wedding gift. The pair, who were then often only in their teens, could be supported by one of them working a score of weeks in the year, and the other a few hours a day. They had nothing else to do.

For centuries mortality during childbirth and infancy had been heavy.

To these causes of a high birth rate were added the tribal fallacy that there is strength in numbers, the protection afforded by a primitive society to a pregnant woman, and the desire to be cared for in old age by one's children. All the traditional pressures led to a massive increase in the Irish birth rate, but after the middle of the seventeenth century more of the new-born survived. There were fewer instances of infant mortality, tribal warfare, murder or mayhem.

It is sobering to plot the increase in the Irish population. In 1660 the population was probably about 500,000. Before the new Protestant monarchy arrived in England in 1688, the Irish population was increased by a generation of prosperity and encouraged by two English kings. Charles II was a secret Roman Catholic, James II an open one. Both were aware of the support which the Irish might give in times of trouble. By 1688, the population had probably more than doubled to 1.25 million. Then the blow fell. The Irish did indeed support James II, who lost to the Protestant King William, and Penal Laws resulted. Though drafted with great nominal legality, they were savage.

Catholics were to be barred from the army, navy, law, commerce and any civic office. They were to be denied the vote. No Catholic could hold office under the crown, nor could he purchase land. Worse, Catholic estates were to be divided and subdivided unless the eldest son became a Protestant, in which case he took all. Monasteries were finally suppressed. Catholic education was made illegal – no Catholic was entitled to keep a school, or attend school, or send his children to be educated abroad. Priests were to be killed, informers bribed and encouraged, and the practice of the Roman Catholic religion proscribed. Ireland became in effect an English colony and her trade and industry went into decline.[19] Urban employment remained at best static, preventing the absorption of the surplus rural poor.

These statutes were later to be used by both the Communists and the Nazis as models of the legal debasement and degradation of a subject people. Such laws reverse the assumption on which civilized society is based – that people tell the truth unless it can be proved that they do not.

What remained of the Catholic aristocracy went abroad to serve in France and Spain. The merchants left, became Protestant, or pretended to become Protestant. Since for centuries merchants have been able to operate in almost any environment, few tears need be shed for them; but to lose the remainder of its ancient, cultured and educated class has been a bitter blow for Ireland. The poor, who are always with us, remained.[20]

Another blow fell on the Irish nation. Until Catholic Emancipation in 1829, that is for five generations, an entire people learnt that the only way to survive was to lie, cheat, dissemble, betray and be cunning as a fox.

Since no justice could be obtained, the peasantry had to create its own. A series of secret and vengeful societies dispensed rough justice in place of the law denied to the Catholics. Deceit triumphed. Guile and charm are all very well, but a complete nation could only survive penal conditions by behaving like an occupied people in a war – not for a few years, but for a century and a half. The lazybed was the European equivalent of living under a coconut tree: no wonder the elegant Englishman of the eighteenth century thought the Irishman lazy and untrustworthy. But he was lazy because he had no regular work, and he was full of guile because the English had passed the Penal Laws, after which his survival was only possible through deceit.

Without the potato and the do-it-yourself cabin, the population could never have survived, let alone increased. There was no possibility of such numbers being fed on bread at the time: Irish agricultural techniques were too primitive to grow the appropriate crops. In the eighteenth century there was no organized grain trade as we know it today.[21] When western Europe had bad weather or disease there was no Canada, no Latin America, no South Africa, no Russia and no Australia to turn to. There was, in general, no international trade to alleviate a local famine, except for an occasional lucky surplus in the young United States.

Between 1760 and 1840 the population in the whole island increased from 1.5 million to 9 million, an increase of 600 per cent in eighty years. Between 1801 and 1841 the population of what is now Eire increased five times. This had nothing to do with the English, but everything to do with the potato. Without the potato all the land in Ireland could at most have enabled only 5 million people to be fed with bread. This was at a period of worldwide shortages of bread grains at a price which the Irish could afford. Compared with wages, grain in Europe as a whole doubled in price in the period 1760–1840. As a result, Irish grain exports were normally the product of between 1 and 2 million acres, say about 1 million tons on average. There were frequent complaints that Irish grain exports increased in step with Irish hunger. This was bound to happen. The weather which gave the English a shortage of food would almost inevitably result in an Irish shortfall as well, and prices rose with shortage, making export worthwhile.

Radical critics of the rulers of Ireland said at the time and still say today that it was a scandal that grain was still being exported while people starved. They forget that the market was common to England, Ireland, Scotland and Wales, and not limited to any one of those countries. More importantly, the starving could never afford grain. The problem of the potato-eaters when their crop failed was not only an absence of potatoes; it was also an absence of almost all cash. This is true of all subsistence economies. Potato-Ireland was merely the last in western Europe, and the Irish cashless society was as shocking to English contemporaries as similar conditions in Africa are to the sophisticated today.

The position of Ireland in 1845–6, the first winter of the Famine, was characteristic – unless such an approaching disaster is recognized while the indigenous population is strong enough to solicit outside help to prevent it, famine is inevitable. But at this stage of the Irish Famine there was insufficient public sympathy to attract assistance, and by the time sympathy had been aroused, it was too late for anyone to prevent the catastrophe. During June and July 1845 it appeared that there was an incipient potato failure. It looked, at first, to be no worse than other failures, no worse than those of 1832 or 1839, though worse than those of 1830, 1835, 1837, 1840 and 1842.

The Irish census of 1851 lists all the noteworthy food shortages, or famines minor and major, local and general, from 1724 to 1849 – that is, for 125 years. It is convenient to divide them into periods of twenty-five years each.

Between 1724 and 1749 there were five failures of the potato crop, of which that of 1739–41 was by far the most severe. Between 1750 and 1774 there were five years of distress – 1756, 1757, 1765, 1766 and 1769. Two of these years were serious enough to qualify as 'famine', and to cause the setting up of relief works on a wide scale and the banning of corn exports. Between 1755 and 1799 there were five years of distress; one, 1784, was of 'famine' proportions – it was mostly Ulster that was affected. Despite these bad harvests, this was a relatively prosperous twenty-five-year period.

Between 1800 and 1824 there were nine years of notable distress, of which five could be classified as 'famine'. In the last of these bad years, 1821, the Irish were universally hungry to the point of starvation south and west of a line from Donegal to Youghal. The government probably did not realize the extent of the damage to the potato, nor the huge size of the population which was by then totally dependent on it for food. During 1821–2 probably 250,000 people died from starvation and allied disease. Between 1825 and 1849 fourteen years out of the twenty-five were classed as 'distress', and there were eight of at least local famine. These years included the great famine of 1845–6.

In 1829 the then Irish and English progressives had pinned all their hopes on Catholic Emancipation, which removed all disabilities from Roman Catholics. But the potato was, as it happened, more important.

Since that date there had not been more than five 'normal' years out of seventeen. To any objective observer, Ireland seemed permanently on the verge of starvation. Half the population depended on the potato for more than three-quarters of their energy requirements. In a 'normal' year, one-third of the population was either hungry or very hungry for at least some part of the year. It was quite clear that Ireland was overpopulated, that the country could only survive if half of the population lived on potatoes and little else, that even then potato husbandry had to be successful, that the weather had to be benign, and that disease had to be avoided or contained.

These conditions were not often fulfilled. As already stated, the population of what is now Eire increased fivefold between 1801 and 1841. Potato husbandry was not universal nor universally successful. The best land was earmarked for grazing or for the culture of wheat, barley, oats or rye, none of which the peasants could grow either on the bogs or in the mountains, and which they certainly could not afford to buy.

Disease was a constant menace, and a new horror would appear at intervals, usually from the New World, as of course did its host, the potato. In the 1750s dry rot appeared. This was what we now know to be a fungus disease, *Fusarium caeruleum*, which affects the tubers in store. Apparently healthy potatoes become dry and shrivelled, ending up woodlike and inedible. In primitive societies the problem was that no one knew why it happened, or how to prevent it. Satisfied peasants would lift their potatoes in October–November, and find that by Christmas they had nothing to eat until the next harvest.

Curl was first reported in the 1770s, and became endemic in the next forty years. This is a virus disease transmitted by aphids, tiny insects which feed on the sap of the plant and transmit disease, in the same way that the mosquito feeds on human blood and transmits malaria. The prevalence of leaf curl is dependent on a native or immigrant population of aphids multiplying quickly enough to spread the virus. The aphid has a limited range, and depends upon winds for migration. While aphids can easily cross the English Channel or the Irish Sea in great numbers, there is no recorded case of an aphid crossing the Atlantic. The virus can reduce the production of potatoes by as much as 70 per cent without any very obvious evidence of anything being wrong with the plant. Natural enemies of the aphid are uncommon; fortunately, because of the prevailing wind, many areas of western Ireland are unaffected by curl, the certain cure for which was not found until World War II.

Botrytis cinerea is a mould which attacks nearly ripe foliage and fruit in a wide range of plants; it first appeared on potatoes in Ireland in 1795. It covers the affected tissue with a blue-grey mould, which sucks the water from the leaf or fruit and makes the affected parts shrunken and dehydrated.

These three problems are bad enough and it is a wonder that, without rotation of crops, hygiene, sanitation or even clean hands any potato anywhere in Ireland escaped to sustain the life of many Irishmen – but two much more devastating plagues were to follow.

Blackleg appeared in 1833; this bacillus poisons the whole plant. By June an affected plant has yellow foliage, with blackened stems and stalks which pull easily away from the main body. Decay of the tubers follows, and spreads during storage, healthy potatoes becoming infected in the ground or in storage. The cure is to use healthy seed, and to dig up and destroy affected plants.

The real killer of potatoes was blight – the fungus *Phytophthora infestans*. It came, presumably, from some dark reservoir of American genetic mischief, and only appeared on the European side of the Atlantic in June 1845 in the Isle of Wight. The disease was reported before 1 August from every country in mainland Europe, and first hit the Irish crops that month, after the earlies were lifted. It reappeared continuously at intervals until microbiologists and agricultural experts found a cure in the 1920s.

Of course, prevention was always known to the successful farmer; to follow the rules of good husbandry, plant only every sixth year, use clean seed in no way connected with previous disease, do not feed diseased tubers to stock without boiling or the disease will be passed on in the manure, and so on. These rules were known to be sensible in relation to all the known diseases of the potato – over twenty in all – in the early nineteenth century. Unfortunately they could not be followed by the Irish peasant, who often had no alternative ground to use, no alternative seed, and no knowledge of hygiene or the means of attaining it.

A bountiful crop may seem to be promised in June, July or August. Suddenly a few plants turn brown and die off. In warm, damp, misty weather blight may spread to a whole field within a week, and within a second week the field will be black and stinking. A characteristic of the fungus is the virulent speed of its spread, perhaps a hundred times quicker than any other pest. In 1845 a man passed through a certain district on his way to Cork for a week's stay with some relatives. On his way south all looked well. On his way back, however, the whole parish was stricken as if by frost, and the fields were black with devastated foliage.

The tubers, if dug, appear to be sound but within a month they go rotten, first with dry rot, then with wet rot. When an area was struck by the blight for the first time the people and the priest thought, with some justification, that God had forsaken them. Perhaps even worse than the physical effect upon the people of Ireland was the idea of loss of luck, abandonment by nature, the fickle quality of the blessings of the Lord.

Nearly a hundred years after the Irish Famine, a learned attempt was made to trace the path which led to the arrival of blight in the Isle of Wight.[22]

The disease had been recognized in the United States in 1843 and had taken two years to reach Europe, perhaps from a potato peeling thrown overboard from an American ship in the Solent or the English Channel.

Blight travelled faster than the next American pest, the Colorado beetle, which had nothing to do with the Famine. This insect took eleven years to cover the United States east of the Rockies and another four years to cross the Atlantic. By 1875 the beetle's success led to a general order in most European countries against the import of American potatoes. The contrast between blight and the Colorado beetle is only mentioned here because of the importance of visibility. The insect could be seen, so some success was achieved in slowing down its advance, notably by arsenical sprays. Because the blight could not be seen, there was no redress against the depredations of the fungus. Nor was there any relief from the peculiar horror of starvation induced by an invisible agent of death. The modern analogy which springs to mind is the effect of radiation sickness.

Free trade as a policy is involved with the Potato Famine in cause and effect, in philosophy and politics, in interaction and interconnection. It is important, therefore, to understand how exceptional free trade has been in human history.

Looked at from today's viewpoint, where virtually no airport is more than a day away from anywhere else on earth, it is obvious how free trade benefits the world, just as internal free trade benefits the United Kingdom or the United States, or France or Germany or Japan. But it was Oliver Cromwell who had to fight for free internal trade in England, Alexander Hamilton in the USA, Napoleon III in France and Bismarck in Germany, and it was not until 1945 that internal free trade was instituted in Japan by the Americans.

Almost no nation really believes in wholesale free trade throughout the world. Everyone wants to buy in the cheapest market and sell in the dearest. Adam Smith, the godfather of free trade, pointed out that goods should be produced in those areas of the world where, together with the cost of bringing them to market, they are the cheapest.[23]

If only one world existed, this would inevitably be true. But if textiles are produced more cheaply in Taiwan because wages are only 20 per cent of the European level, or motorcycles in Japan because the culture of the Japanese lends itself to mass-production, or ships in Korea because Koreans enjoy the heavy capital investment of the Americans and a lower wage level than any other country involved in shipbuilding, who is to believe in free trade? In the long run, free trade benefits everyone; in the short run it is bound to produce much pain. Furthermore, free trade benefits the strong, the competent, the people who use new technology more than those who are weak, either permanently or temporarily, managerially incompetent, or those who produce by means of ancient ways.

Since about 1780 free trade has been seen by a few people as an ideal at which we should aim, by a few more as pragmatically possible. The latter say that if it pays, let's adopt it in whole or in part, as long as it is right for us. The vast majority are, were, and always have been against free trade.

If this is felt to be too extreme, it is worth pursuing the point. Free trade always seems to imply, first, free trade in consumer goods, but other consequences very rapidly follow. If imports are cheaper than the home-produced article, either the price of and profits from home-produced goods must be lowered, or the home-produced factory and the producers both become idle, redundant, unemployed. Both capital and workers suffer. Capital goes abroad to find a better return and to balance the cost of imports. Workers emigrate, stay unemployed, or change to another industry. Europe and the USA have seen much of this process in the last ten or twenty years. Production of vehicles, ships, textiles and electronics has gone to the Pacific if subject to free trade; European and American industries have suffered, some very severely.

The argument about protection – if that is the antithesis of free trade – of people's food has always been more heated than discussion about cotton or wool or china, to mention three trades about which there was much British activity and legislation during the eighteenth century. Whenever at some point in history a well-meaning do-gooder has sought to change a people's staple from chestnuts to rice, from rice to wheat, or from wheat to maize, there has been trouble. Each alteration of diet, especially an alteration involving a foreign supplier, has been seen to be a change for the worse. Ample food in a world which was generally short of food meant political and economic power. Before steam it also literally meant power, since apart from wind or river, there was only manpower (= food) or animal power (= food). Even metals could be expressed as costing so much in terms of food. Food was of obsessional importance to the whole world before the invention of twentieth-century techniques which dramatically reduced the ratio between the cost of food and the wages paid per hour. In some countries food is still power, still scarce, still an object of unceasing concern.

Food was also always a political matter because of food-producers. In no country in the world before 1850 was less than half the population engaged in food and agriculture, raw, processed and manufactured, ancilliaries included. The production, distribution and exchange of food are of primary importance, and without them all other activity would grind to a halt. The more primary the producer, the more worthy the politician has felt him to be, especially if producers were numerous and articulate.

Even before they had the vote, peasants were of great importance as workers, as soldiers and as serfs. The countryside was a reservoir and a breeding ground for the diseased towns until the success of sanitation in the nineteenth century. Rural areas were seen more recently as a pool of underemployed people in slumps, to be urbanized during booms and so forth. Every continental ruler in the nineteenth century regarded the peasantry in much the same way as his ancestors had regarded the serfs – a positive reserve of simple, productive people enjoying a useful, peaceful life until required for something more urgent, or unpleasant, in peace or war. This perspective never applied in Ireland.

The peasantry were not, as in the rest of western Europe, in communion with their lords. The attachment of slaves to lords is a completely different process, as has been explained, from the attachment of both to the land. If both serf and master are attached to the land, both are serving an earthly object upon which they can agree. If one is attached to the other, the result is the equivalent of today's bosses and workers, with all the conflict which that relationship implies.

It is also true that the English Ascendancy in Ireland sought its fortune in another country. By the eighteenth century there was not much difference between the rich, high-born and famous in Ireland pursuing their political future in London, and the equivalent French men and women seeking wealth and success in Versailles. Not for either was the organic unity of the countryside. Even the humblest English squire was happier in rural Norfolk or Shropshire than his plantation equivalent in Ireland. Rarely did an English or Scottish Protestant feel truly at home, except in parts of Ulster. The Emerald Isle was a source of income, a country to be exploited

The English turned Ireland into an agricultural colony, and prevented the manufacture of anything which they, the English, felt should be produced in England. This policy has been castigated by many modern Irish and their intellectual allies as 'jealousy'. In fact, it exactly accorded with the whole of post-Renaissance mercantile theory, and was imposed by all Europeans upon their colonies and, wherever possible, by the English in both America and India. Ireland was one of the great successes for English mercantile theory. Proximity and the helplessness of Ireland were more the causes of English success than any amount of 'jealousy'.

The silk trade was destroyed in the early 1700s. Cotton imports from all over the world were choked off in 1722. Colonial brewing was prevented by the prohibition of the import of hops in 1731; Ireland had to brew porter, not beer.[24] The export of glass was prohibited from Ireland in 1765. The fine linen trade was destroyed by the prohibition of the manufacture of cambric and lawn in 1767. The direct import of sugar, tobacco and even spices was prohibited from the 1690s.

The relationship of Ireland to Great Britain in the period after 1660 was therefore that of a colony in a mercantilist scheme. There were two intermissions: the first is not well known, but had a great influence on the ultimate disaster. After 1665 the Irish were denied the right to export cattle, alive or dead, so the graziers naturally turned to sheep. Then they were denied the right to export wool, to spin it or weave it. The previously sheep-free pastures produced better wool than any from sheep-sick, overgrazed mainland Europe or England. So the wool became a high-value product worth in France nearly twice as much as off the sheep in Ireland. Wool was smuggled in huge tonnages. The smuggling meant bribery, connivance by the rest of the population and corruption. By 1695 more illegal European goods were being imported into Ireland to pay for the wool than the legal landings in the ports, which, by law, all had to come from England. The disrespect for the law spread from the Penal Laws against Roman Catholicism to all laws concerning the English Ascendancy, religious, economic or political. After all, they were manifestly unfair. Communal lawlessness became a national defence mechanism; disrespect for the law of man a necessary way of life; a dreadful 'justification' for the violence to come.

A great deal of attention is drawn by the Irish to the second break from colonial status. This occurred when Pitt the Younger was Prime Minister of England, and is known as 'Grattan's Parliament'. Henry Grattan was an orator in the very finest rhetorical tradition, later called 'Ireland's Demosthenes' by the Whig leader Charles James Fox. He became leader of the national party in the Irish Parliament in 1780 at a time when it was becoming obvious to the English that they were going to have to admit that they had lost the American colonies. Two years later, just after the British surrender at Yorktown, the Irish Volunteers held a huge convention at Dungannon.[25] The British government gave in and repealed Poyning's Act, a law of Henry VII, which had laid down that all proposed Irish legislation had to be approved by the English Privy Council before submission to the Irish Parliament. It could then only be approved or disapproved by the Irish, and not amended. This power had been reinforced at later dates, so that the Irish Parliament was totally subservient to the wishes of the English establishment.

In 1782 the Irish wanted independence, but they were not in revolt. 'I found', said Grattan, 'Ireland on her knees. I watched over her with paternal solicitude. I have traced her progress from injuries to arms, and from arms to liberty. Spirit of Swift, spirit of Molyneux, your genius has prevailed. Ireland is now a nation.' When he uttered these words Grattan, self-styled father of his country, was only thirty-eight. The British, however, were influenced more by the 100,000 volunteers who were organized and armed than by Grattan's brave words. Nevertheless, lest it be mocked, Grattan's Parliament immediately raised taxes amounting to over £250,000 for the Royal Navy; they were not disloyal. Like the early American Revolutionaries in the 1760s, they wanted to be taken seriously.

For sixteen years there was an era of great good feeling. There was promise of Catholic Emancipation. There was promise of free trade between England and Ireland. There were promises of reform of the franchise, of commutation of tithes, of religious toleration. Although nothing came of any of these promises, the good feeling survived.

What brought both the good feeling and the independent Irish Parliament to an end was the war with Revolutionary France. Following the precept that 'England's danger is Ireland's opportunity', and imbued with French Revolutionary ideas, Wolfe Tone and his United Irishmen rose in revolt in 1798.[26] Tone was a Protestant, a socialist and an anti-clerical. The rebellion was put down with brutality. A movement for union with the United Kingdom was promoted with much bribery by the British government. It was supported by the Roman Catholics, seduced by a promise of Catholic Emancipation in the Westminster Parliament. Union was opposed by Orangemen and Episcopalians.[27] It was carried by means of the worst type of politics, but George III would not allow Catholic Emancipation. Pitt resigned.

Hardly anyone noticed the most important factor in Ireland during Grattan's Parliament. During the eighteen years before 1801, the population increased by an almost unbelievable 90 per cent from 2.6 million to just under 5 million. During the first, hungry years of the nineteenth century there came a check: the increase in 1801–21 was 'only' about 1.9 million, or about 37 per cent; this period included eight years of famine.

Trade and commerce were under the full control of the English, and any attempt to set up new industries in the towns on the English model was discouraged after the Act of Union in 1801. Dublin became down-at-heel, residential with a minimum of trade, commerce or industry; its chief occupation was rhetoric. Irish towns, including Belfast, could not absorb the rural unemployed as did towns in every other country that was becoming industrialized. Ireland was not, however, becoming a workshop. It was no more industrialized in 1840 than the Ukraine, and much less a manufacturing power than India or China.

Land, however poor and inadequate, was a necessary meal ticket, and the rent for that land was double the equivalent in England. There was no alternative, little work and little money, no relief, and no soup kitchen outside the workhouse. And the cash went, ultimately, to the landlord, of whom only a very few lived on their estates in Ireland, knew their tenants and treated them as human beings rather than as a source of income.

In the 1830s the Great Liberator, Daniel O'Connell, compiled a list of 'good' landlords: they owned less than 5 per cent of the land area of Ireland. The average Irish landlord was neither an Irishman nor a Catholic. He was a member of the Protestant Ascendancy, an absentee in Dublin, London, Paris or Rome. More than likely in debt, he bled his tenantry for the interest on a debt incurred by himself, his father or grandfather in aggrandizement of self, house or political career.

At least £5 million a year, every year, went abroad in rents to sustain the lifestyle of absentee landlords. This sum, plus the import costs of English manufacturers and tropical products like sugar and tea, another £10 million at least,[28] led to a permanent financial deficit in Ireland, and weakened the base of any patriotic class of men of substance.

For four or five years after the first attack of the blight in 1845, Ireland was racked by the physical agony of famine, disease and depopulation. Although 1845–6 is usually regarded as the span of the Famine, the blight in fact struck several times more, though not as severely as in 1845; and the after-effects, in particular emigration, continued for a long time. Up to a million men, women and children are estimated to have died from starvation, or typhus, or cholera, or one of the other virulent diseases that followed in the Famine's wake. Additionally, up to 1.5 million Irish left the country as a direct result of the Famine, setting a pattern of continuing emigration for the rest of the century; by World War I some 5.5 million people would have left Ireland. The history of both Britain and the United States was fundamentally altered by this situation.

This was the worst recorded famine to take place in Ireland because, though there had been no fewer than twenty-seven famines in the previous hundred years and five in the preceding decade, and their frequency was accelerating, these other famines were all local. The disaster of 1845 was not only the first recorded island-wide potato failure, but from some point in the autumn of that year, for the first time in history, there were no potatoes for sale anywhere in Europe – for two years virtually none was fit for market.

The dearth of potatoes placed an immense strain on all other staple starchy foods. The price of wheat rose from less than £13 per ton in July 1845 to more than £30, and then fell back to about £10 per ton in September 1847, when the Famine was over. The equivalent in terms of today's money would be £1300, £3000 and £1000 respectively. As a comparison, imported 'hard' American or Canadian wheat costs about £130 per ton today in Europe; a price of £300 per ton, equivalent to one-tenth the highest price paid during 1845–7, would rule out most bread-eating. In the 1980s people would merely switch to rice, rye, oats, maize or barley, but in the two years of potato blight there was a worldwide shortage of grains, and inevitably what supplies there were went to the rich. The Irish were not only poor, but more than half the population never saw money for most of their lives. They were not within the cash economy.

It is the infrastructure of a country which prevents famine, and its total absence which makes alleviation very difficult, if not impossible. Examples of this truth, such as Ethiopia and Sudan, abound in the 1980s. The Potato Famine was the last great famine in Europe, because the improvement in infrastructure made such disasters less and less likely. In the future any famine would be as the result of the breakdown of the normal structure of life, such as in or after a war, political change or economic folly.

There was no telegraph between England and Ireland until the 1850s, and no transatlantic cable until the 1860s. The lack of these means of communication made much more difference than the actual transport of the grain from places of surplus to places of shortage. Until market intelligence became available on a worldwide basis, at some point shortly after the American Civil War, no forecast of disaster could ever be made. World weather patterns are very variable, and in 1845 there was no proper appreciation of conditions in the New World or the Southern Hemisphere. It was not economically viable to prepare for a shortage which might or might not take place. Furthermore, grain storage at that time, which was mostly in stacks or ricks, was a doubtful proposition; off-farm storage was not safe from pests or weather until concrete or steel silos were introduced in the 1880s. Nor can a market be self-managing in the absence of reliable statistics. For all the strength and power of steam railroads and iron ships in early Victorian times, in the absence of good, fast communications the power to prevent or avoid famine did not exist.

There had been a famine a century before – in 1739–41. No records were kept of the numbers involved, but it seems probable that the proportionate losses were higher – perhaps a third of Ireland's 1739 population – and for the same reason: reliance on the potato. The population, as far as we can estimate, was probably about 1.5 million at the beginning of the 1739 famine. Half a million deaths is a not inconsiderable horror.

The difference between the 1739–41 and 1845–6 famines was that very few people, if any, knew about the events of the earlier catastrophe. If there had been newspapers, eye-witness accounts and so forth in 1739–41 the famine might have taken its place by the side of the new war against Spain. Almost everyone in England had heard of the *casus belli*, Jenkins' ear, cut off by the Spaniards and exhibited to a shocked House of Commons. Hardly anyone knew of the famine that was taking place in Ireland at the same time, and very few are aware of it even today.[29]

So the disaster of 1845–6 owed a great deal in its importance to the existence of newspapers. This is not to be callous or to pretend that it was a media event; but about 40,000 London papers were sold daily during the period of the famine. More than half this circulation was that of the thundering *Times*, radical in those days, which did not allow the well-fed, comfortable middle classes to ignore Ireland or its troubles. Every day, as the files testify, the miseries were recorded.

The English middle classes did not in fact wish to be complacent, but at that time the concept of laissez-faire freedom was very popular. It was the belief of the age that, just as physical shortages could be changed by increased freedom to produce, to distribute and to trade, so too could political problems be solved by more freedom. The Reform Act of 1832 had given power to the middle classes, who were reforming, nearly always Christian, often conscience-stricken, sometimes busybodies in their desire to help. (Marx and Engels had not yet pointed out to the middle classes how incurably bourgeois they were, and fortunately they had no opportunity to realize this deficiency for themselves.) The belief was therefore strongly held that all problems could be solved, if only enough freedom existed in the marketplace.

A little time spent considering the physical facts of the Irish problem would have produced an argument along these lines. The country is grossly overpopulated because potato culture can support a higher population in any given area of agricultural land; emigration, recovery of waste land and modern methods of cereal growing must be encouraged; a cash economy must be established so that the cashless society of the 'truck', involving perhaps 4 million people, can be brought to an end. As an immediate remedy for the crisis, all exports of all grains from Ireland must be stopped; the exporters must be compensated for their loss; soup kitchens must be set up all over the potato districts to help feed the starving. Unemployment relief measures must be instituted, as in England. Money must be poured into Ireland to pay for these measures, but value must be obtained for the English money so invested.

These provisions would not have amounted, in all, to more than £60–75 million sterling, about ten times the cost of the Great Western Railway from London to Bristol a few years previously. For this sum the Irish nation could have been saved, Ireland industrialized, and the peasants moved forward from the sixteenth to the nineteenth century. In the laissez-faire climate of Britain in 1845, however, these measures were not likely ever to have been taken.

Yet all the initiatives mentioned above had previously been adopted at some time or another in England or Ireland, and it was within the abilities of the English, if they had cared enough, to repeat them. It is not unreasonable to share the Irish peasants' loathing of the English reaction to their problems. They wanted food; the English gave them words. They wanted work; they got eviction. Ireland needed investment; the island was told to look after its own poor. Disease followed dearth, as night followed day, and soon, in some areas, there were not enough fit men to bury the dead.

Some people wonder now, as others wondered in 1845, at the contemporary inability to see what was needed and what could have been provided. How much of the Irish tragedy was the result of what the Jesuits call 'invincible ignorance'?

In 1845 some of the British could be said to be selfish, opinionated and determinedly obtuse, if not invincibly ignorant, about the Irish. If this is fair, it was the fault of the success of the theories of the English mathematician Thomas Robert Malthus (1766–1834). His father was a friend and admirer of the French philosopher Jean-Jacques Rousseau, and subsequently his executor. When Rousseau died in 1778 Malthus senior asked his thirteen-year-old son to read and admire some sentimental and unpublished writing of Rousseau's which reinforced the proposition that mankind was perfectible. Such an experience in adolescence might well mark a man's intellect for life.

The contrast between salvation on this earth and redemption in the next has occupied philosophers of every serious religion for at least three thousand years. When Malthus junior reached Cambridge as an undergraduate in 1784, he wrote his father a stiff note about man's inability to improve himself, presenting the opposite view from Rousseau's. Though the older man disagreed with his son's argument, he admired it. Each encouraged the other, and they were well matched. Father and son carried on a correspondence for the whole of the son's time at Cambridge, their love and respect for each other in no way diminished because they so profoundly disagreed on fundamentals. In a way, they represented the two strands of thought, conservative and liberal, which have been the warp and woof of every democratic society: welfare versus growth, hard money against soft, laissez-faire or intervention, free trade or managed systems, and so forth.

Young Malthus was confirmed in his opposition to the Encyclopaedists – Condorcet, Rousseau and Diderot – by the absurd sentimentalism with which the French Revolution was greeted by the sort of people who agreed with his father. During the first phase, 1789–93, the English 'progressives', led by politicians such as Fox and Grey, were so indiscreetly and impartially favourable to the Revolutionary cause that they made themselves suspect and their cause impotent for more than a generation. They also seriously delayed reform, because they confused liberty with licence. Pitt, the reforming First Lord of the Treasury, had to drop most of the measures in which he believed; he became a reluctant reactionary. It suddenly became respectable – fashionable even – to support self-interest. The change in attitude was analogous to that of the 1980 American Presidential Election when Reagan defeated Carter.

Fourteen years after he went up to Cambridge as an undergraduate Malthus published the first edition of his *Essay on Population*, which met a philosophical need. The problem about declaring that it is impossible to assume the perfectibility of man is that a voice from somewhere will shout: 'Prove it!' For a whole generation and more, Malthus' doctrine 'proved' that mankind could never be perfect. Malthus made the proof a matter of 'scientific fact', not sentimental opinion. Believers in the status quo were much relieved.[30]

The basis of Malthus' theory is simple. Man can never be perfect, because hunger will always be the limiting factor. Population increases by means of a geometrical progression, while food increases by an arithmetical ratio.[31] This statement may well be false, since it takes no account of technological improvement in food production and does not allow for the limiting factors in population growth. But in 1800, and for two generations afterwards, it was intellectually respectable to believe in the proposition. And why do people breed? Because they are carnal and vicious. This also could be believed.

In fact, before the nutrients necessary for plant growth had been identified, and at a time when the only plant foods were animal or plant residues, and when oxen, men and horses were the only power (and they themselves needed food), and when a typical cow or steer would not come into profit before the animal was four years old, hunger was a common enough occurrence in England, let alone Ireland. England had the best infrastructure in the world in 1797, Ireland one of the worst in Europe. Hunger was not only a question of growing food, but more importantly a question of harvest, distribution and exchange. The Irish peasant grew his own food, and was outside the network of distribution and cash exchange. What would happen if the potato crop failed?

Between 1797, when Malthus' book first appeared, and 1845, there were twenty failures of the potato crop, all of them involving some deaths from starvation, disease or debility; all of these years would be classed today as famines. In twenty years out of forty-eight there were famines in Ireland, but only three in England. The Irish population was increasing at least twice as fast as the English. They were regarded as carnal and vicious.

It was easy to believe in Malthus – and people have always believed what they wanted to believe. Belief was subsequently made even easier by Malthus' disarming modifications of his theory. He qualified his claims: it was not original, it was not absolute. Population *tended* to increase faster than food supply. Others had not neglected it: the problem was mentioned by philosophers and economists such as David Hume, Robert Wallace, Adam Smith, Richard Price, Montesquieu and Arthur Young. The *Essay on Population* ran into six editions in Malthus' lifetime, each more restrained than the last. No matter: those who wanted to reinforce their prejudices, and do nothing about Ireland, could read and quote the most powerful Malthusian case, as expressed in the first edition of 1797. The Irish, so the theory ran, were hopeless; they were incurably vicious. They would always breed faster than the food supply. The only control over population growth was the occasional famine, which they deserved anyway because they were so vicious. . . .

It is important to realize the long-term – if crude – truth of Malthus. It could indeed be proved that without the oil industry, food would never have caught up with population. Hunger would be general without fertilizers, weedkillers, pest control, infinitely variable feedingstuffs and mechanization, all of which are available only courtesy of the oil industry. There was no hint of the existence of a 'modern' petrochemical industry such as we now enjoy until the end of World War II. Significantly, the possibility of preventing world hunger has only existed since 1950.[32]

It was the combination of mathematical logic and high moral tone which gave the works of Malthus their ethical power. It was the same combination which enlightened the arguments of the opposition to Malthus' pessimism. This was the early Victorian (Albertine) faith in the efficacy of free trade. It was not wholly convincing. There were two reasons, both connected with the Corn Laws and the Anti-Corn Law League.

Innumerable Corn Laws had been passed in England since 1436, covering the export or import of wheat and barley from England and Wales. Trade from and to Scotland and Ireland and the colonies was given preferential treatment, but was subject to duty or bounty. The intention of the laws – and every other European country also had statutes of a similar nature – was to try to keep the price of wheat steady, so that if prices rose through dearth, the import duties could be removed; in the event of a glut, bounties or subsidies were applied to encourage export.

By the beginning of the nineteenth century, the laws had become very unpopular in certain quarters. The state of balance that had brought the laws into being was no longer working because Britain was importing so much of her corn supplies: by 1840, 25–30 per cent. The Napoleonic Wars had brought about inflation and the restriction of imports, driving up prices and the rents for agricultural land. In 1815 and 1822, import duties were raised in order to maintain prices, but in the postwar period deflation led to recurring crises of unemployment and falling prices, and there was an outcry against high bread prices. The free trade lobby, which objected to the laws in principle, was growing more vociferous, and to this vociferousness was added political influence with the passing of the Reform Act of 1832. Manufacturers and industrialists wanted in particular to remove the duty on grain in order to secure cheaper food for their workers, as well as cheaper raw materials and an expansion of their markets.

Agitation against the Corn Laws centred on Manchester under the political leadership of Richard Cobden and John Bright, both cotton manufacturers. In 1839 they founded the Anti-Corn Law League, and fought a clever and effective propaganda campaign against the landowners, most of whom were staunch Tories. In 1845 the English harvest was short, while Ireland was in the throes of the Potato Famine. Urged on by the League, Sir Robert Peel, the Tory Prime Minister, decided to turn from protection to repeal, reducing the wheat duties in June 1846 and abolishing them as from 1849.[33]

But, despite the persuasive arguments of Cobden and Bright and the political turnabout of Peel, it should have been obvious to all that mere free trade was in no way going to help Ireland. The shortage was worldwide, and it was of food. No amount of repeal of the Corn Laws was going to produce one extra sack of grain or give the Irish peasantry the money they never possessed, for there was no surplus corn anywhere in the world.

Secondly, the Anti-Corn Law League had never mentioned Ireland in all their propaganda, a point that was made little of, except by the League's implacable opponents. No one had imagined that Ireland's misery could be used to destroy the Corn Laws in England. Repeal would have a very marginal effect in Wales or Scotland, and hardly any in Ireland, yet it was the misery of the potato-eating classes in Ireland which led the Albertine conscience to repeal. The *Economist* was a weekly newspaper founded almost wholly to support free trade. In its first year every edition published a calculation of what the Corn Laws had cost the country in the previous week. In 1844 there was not a word about Ireland, potatoes or the benefits to the Irish peasantry of free trade.

In fact, of course, the English establishment comprised neither fanatical Malthusians nor fanatical free traders. They were concerned that the Queen's government must be carried on. They were concerned not necessarily to find out the truth about Ireland, but to limit the effects of a catastrophe. Practical men very often have to deny themselves the joy of searching for the truth, whatever that might be.

In order to 'cure' what he thought was a temporary problem, Peel imposed upon England a pattern which could never have benefited the nation as much as the various alternatives. But did Peel expect to be defeated in a subsequent election? Did he not hope to carry his party with him? Did he expect to be killed by a fall from his horse in 1850? If Disraeli had not whipped up the irreconcilables would not a compromise have been possible? We can never know. But the damage was done. Telling a highly emotional, sentimental and illogical middle class that free trade was in the same group of good causes as those favoured by Charles Dickens was to perpetuate trouble.[34]

Free trade, incidentally, was only a moral issue at the ports. Internally, there were as many make-work provisions as ever, as life became more complicated. Learned societies, professional bodies, authorities, examination boards, the Civil Service and so on made Britain an example for the world of middle-class restrictive practices, and made the professional associations as mercenary as any trade union. It is now obvious that free trade should have been bargained against tariff reductions in other countries, as it is today. Free trade by only one country was to prove the precondition for British industrial decline. It should be remembered that few other countries followed the British lead.

The second point, whch arises out of the first, was the concentration on food free trade. This inevitably turned the United Kingdom into a low-cost labour country, since any manufacturer's labour costs could not exceed those of his competitors. Free trade meant that only the brighter people emigrated, leaving at home the relatively less well educated, unskilled and low-paid, who were unwanted abroad. It meant that the UK was set in a pattern where its prosperity depended on low wages and not on capital investment, know-how or skill. Almost of necessity free trade meant exporting capital to pay for imported food, or to give the customers for British manufactured goods the money to pay for the cotton goods, coal, iron and steel which the British had to export – as an alternative or additional to capital – in order to pay for the food they could have grown at home. Britain also became, by definition, the greatest trading power the world has ever known. This meant her near-defeat in two world wars because of an economy which could not function without the import of 20–25 million tons of food and raw material a year.

Both world wars would have been more easily won by Britain if the tonnages needed had been reduced. Indeed, the rivalry with Germany might well never have occurred if the trade/navy dominance of Britain had not become an absolute necessity. Imported food became paramount, and became as well an unexpected but contributory cause of World War I. This was partly because the opportunities for export to developing countries diminished as those countries developed their own manufactures. So a large captive zone of customers became a necessity, of which the largest in numbers of people was India. The results of the free trade Empire policy were sometimes surprising. The Malayan boom in rubber in 1910–13, for example, had an appreciable effect on Britain exports because, while Britain only consumed 25 per cent of Malayan rubber, the management of the whole industry was in British hands.

The trading empire led to an immensely complicated interlocking service industry involving shipping, insurance, banking, telegraphs, and all sorts of provision of permanent capital; even the Victorian public schools to produce reliable young men to administer the natives. No one at the time had the necessary data to analyze the system, though many admired it. Now that it can be analyzed, admiration is muted. It is easy to conceive a policy for the British which, starting with the basis of not making any concessions without receiving any concessions in return, would not have had the long-term negative results which did occur. In 1959 Dean Acheson said: 'The British have lost an Empire and have not yet found a role.' Britain ought never to have acquired the post-1845 commercial empire in the way it did: it was not inevitable. Without the Irish Famine, it would not have happened so quickly and so unfavourably as it did.

The worst effect of Repeal, however, and an ironic one, was the production of what Disraeli had already warned against, but for different reasons. Free trade produced two nations. Wages had to be 'contained' in order to export cotton piece goods (90 per cent of world exports in 1845) or iron rails (70 per cent of world exports in 1845), or coal (65 per cent of world exports in 1845). Cheap food would not only do this, but would also permit the substitution of unskilled labour in any industry if the job were kept unskilled and the pool of unemployed was retained. The British were driven by cheap food into perpetuating a poorly paid working class. This effect was the reverse of the golden future that had been promised by Bright and Cobden.

To remain competitive, an exporting country must have significant advantages, but they need not always comprise low wages. Eighteenth-century British North America (modern Canada or the USA) could build wooden ships more cheaply than anyone in the world because, while North American workers were well rewarded, they were often self-employed and always canny, and raw material was relatively very cheap. Before the industrialization of cotton, India could produce the cheapest ethnic, village fabrics in the world, sometimes of great beauty. In order to raise the earnings of cotton piece-workers before the Industrial Revolution, the British government raised both tariff and non-tariff barriers against India.[35] The same was true of Chinese porcelain. The English consumer was deprived of both products. An open-cast deposit of coal (a 'drift' mine) could be made to produce cheaper fuel in the period before pumps made deep mining possible, and again in modern times after the development of earth-moving machinery. For many years before World War II coolies in China, on very low disposable income, stripped land and removed the coal like a lot of ants. This sort of coal could not be profitably produced in Europe or America at the time; it can today, anywhere. This coal, like European deep-mined coal in the nineteenth century, competed because of low wages.[36]

In a low-wage economy like that of the UK after 1850 there was no incentive to improve performance by all the other non-wage methods of management. Technique, know-how and imagination frequently cost very little more in money wages than is paid to people without these qualities. Good managers should choose systems for their cost-effect, not merely for their low input cost. High-wage production *can* work. Examples are the Swiss textile industry, French wine, Italian electrical goods, Scotch malt whisky, German cars, Dutch vegetables and Israeli citrus fruits.

All of these industries prosper because brains are more important than brawn. All illustrate the power of challenge and response, all of them break economic rules, all of them prove the perverse human ability to triumph over adversity. It was in this sort of activity – requiring non-purchasable human attributes – that the English declined after 1850. Only now, in very different circumstances, are the activities requiring technique, know-how and imagination beginning to flourish again in England: tourism, design, the media, financial services without the dominance of the City of London.

In America, the Irish influence has become so much a part of political life as to be taken for granted, but before the 1840s the whites in the new nation were overwhelmingly Protestant, Nordic or Anglo-Saxon, from England, Scotland, Wales, Ulster, the Netherlands, Germany and Scandinavia, with a few pockets of Catholic French, German, Spanish and Swiss. There had been a minority of indentured servants, beggars, vagrants, minor thieves and prostitutes shipped out to Virginia in the seventeenth century, and all over the Middle Colonies and the South in the eighteenth century before 1776, and a proportion of these would have been Irish and Catholic, from both Ireland itself and the slums of English cities. Most of the 'Irish' immigrants in the eighteenth century were from Ulster and Protestant. Fewer than 100,000 Irish Catholics entered the USA between 1800 and 1840; their numbers were not of great consequence, and above all they did not form a cohesive bloc representing the same viewpoint or sharing the same ethnic origin.

The flood of Catholic Irish began in the 1840s. At the beginning of the Famine immigration of the destitute was discouraged, as it had been for fifty years – there was a method of control involving minimum fares on ships from Europe. Before 1847 many of the Irish who entered the USA walked across the Canadian frontier. But once the Famine was written about, considerations of humanity overcame public policy. In the fifteen years between the Famine and the American Civil War Irish incomers reached a total of more than 100,000 a year. They made a difference to the new nation not only in terms of numbers (about 5 per cent of the population of those states which they entered), but also as a proportion of the favoured cities (about 30 per cent). Boston and New York were often the first places the steerage passenger would see, and there he often stayed.[37]

The immigrants were usually destitute, often diseased, unskilled labourers, without homes to go to, and with few friends and relations to help them. They had, perhaps, been the survivors of several years of starvation, certainly of several years of protein shortage, and exposure to typhus or pneumonia. They huddled together in Irish Catholic ghettos; the priest was their friend on Sunday, and the local politician their amiable exploiter during the week. This group of potential voters could be massaged into blocs of great political power, which still influence the cities that first sheltered them. Out of the subculture which they formed came, in the next two generations, rich and powerful politicians and property men, and their Celtic qualities were in no way diminished by this significant urbanization of the Irish Catholic character.

While English cities, and Glasgow, were similarly populated by immigrant Celts, these people, unless they were householders, had no vote. They never felt part of their adopted city because they were never more than a few days' journey away from home. In America, on the other hand, the Irish peasant, who had grimly endured the whole previous generation, became a legitimate citizen whose vote was canvassed, whose needs were met and whose importance was freely acknowledged. The flowering probably took fifty years – it took time to overcome the appalling conditions which had driven the emigrants to leave the bogs and mountains, but a new and amazing vigour and enterprise was evident in every industry and service into which they entered: railroads, textiles, mining, contracting, building, civil engineering and the police, as well as politics.

The art of talking the hind leg off a donkey was given full rein, and wordy rhetoric, so characteristic of the public man in the later nineteenth century, became more fashionable than anywhere else in those cities with a large Irish Catholic element. The native Irish enjoyed it all, whether the rhetoric was their own or that of some politician of another ethnic group soliciting their support. On the other hand Celtic mistrust of the overlord, which had been a way of life and a means of survival for the Irish for nearly two hundred years, did not disappear at once in the new conditions, and the Irish heightened political volatility and strengthened the American belief that any action is justified by success.

Of all the characteristics which the Irish brought to America, there was one which the Founding Fathers might have found disturbing if they had lived to see the 1850s. Though nearly all immigrants to America, from the fortune hunters of the seventeenth century onwards, had some compelling *negative* reason for leaving Europe, nevertheless when fortune smiled upon them in the new land they came to terms with the old country, traded with England, borrowed money from London, accepted the protection of the Royal Navy, sent their children to be educated in the Old World and imitated the best of European manners and artifacts. Of all the immigrants into the United States at a time when a high proportion of Americans were of Anglo-Saxon Protestant origin, the Irish were the only politically organized, ethnically integrated group motivated by hatred for England. Though George Washington, Thomas Jefferson and others were not professional WASPs, and though they discounted the value and importance of Europe, they were in essence provincial, transatlantic Englishmen. Trade, credit and communication, language, literature and law all conspired to make the Anglo-American cousinhood a mutually rewarding relationship, with no more than normal cousinly friction.

Irish Catholics were the first group to bring hatred into Anglo-American relations and the first group to adopt deliberate amnesia about the errors in their own past. There was every reason for these unhappy refugees to hate England: cold-hearted, logical, pessimistic England, which had repealed the Corn Laws but not produced any more food to eat. This Irish hatred of the Anglo-Saxon also appreciably sharpened the attitudes in Northern cities towards the WASPs in the South. Many of the Irish voted in 1860 for hard-line Republicans determined to diminish Southern power, to destroy Southern pretensions. The Irish were more anti-English than anti-slavery.

The potato monoculture caused the disease which caused the famine which caused the emigration to America. Arrived in the USA, the Irish established a pattern in the cities which other ethnic groups followed, and edged the new nation further into a sense of difference from Europe. Before Zionism, the Irish lobby was the strongest in the nation; Irish fund-raising, both legitimate and illegitimate, the most effective; Irish denigration of British imperialism the most articulate. The Irish turned America into a champion of the anti-imperialist school, delayed American entry into both world wars, and continue to blacken British political and diplomatic effort, however easy that may be to do.

In Ireland itself, the tribal struggles of a seventeenth-century ideological inconsequence continue. The divided island is an example for anyone who would like to ponder on what post-Reformation Europe would be like without the safety valve of emigration. The tribal war continues in Ireland, the numbers of people involved increased because of the potato, the history made more bitter, the folk memory on both sides sharpened.

Did the Potato Famine affect more than just *Irish* numbers? If it had not been for the Famine emigration to America, is it not probable that the United States would have remained largely WASP? Would Italians, Jews, Russians, Poles and other continental immigrants have been allowed entry in such numbers? The United States and the United Kingdom, with similar populations of 22 million and 18 million respectively in 1844, would have marched down a more similar path, the United States leading the way, but more comparable than has been the case during the last 150 years.

Affirmative answers to such questions, while producing a duller United States and a much duller New York City, might have prevented both world wars and all their significant consequences. This is one of the most pregnant of all of the 'ifs' of history, and the least regarded. But perhaps the potato has done enough, in all conscience, without examining the more extreme possibilities.

Notes

1   The potato propagates underground by tubers, each 'eye' of which is notionally capable of producing a new plant, or by ordinary overground seed, which is not always produced in modern varieties. Wild South American types flower and set seed in the ordinary way, and it was this seed which was originally brought to England, as well as tubers. Edible underground roots were at this date rare and it was the seed of most plants that people ate, so Raleigh's actions were not that odd.

2   The name 'potato' can cause as much confusion as the history of the vegetable's arrival in Europe. Until 1770 three plants were known as 'potato': the white, which is the subject of this chapter; the yam; and the sweet potato. All are useful and edible, and the last two were available in the Caribbean as ship's stores from about 1550 onwards. There they were locally called *battata*, and after 1600 all three vegetables were called potato, *patata, potaton, potade* and so forth. A variant of this name existed in every European language until *pomme de terre* and *Kartoffel* arrived.

3   The mountains of what is now County Wexford are visible from the hills of the extreme southwest of Wales, 58 miles away. There is a pre-Celtic ley line from a point near Brunel's railway excavations, above Fishguard, to a point near Rosslare; the two ends of the line are clearly visible. (Neatly enough, in view of the subject of this chapter, Brunel's railway workings were abandoned because of the fall-off in traffic across the Irish Sea due to the potato famine in 1845–6, and the railway was subsequently built more economically via a different approach to Fishguard.) There is no reason, except fear of the weather, why Wales and Ireland should not have been in contact. There was a great deal of pre-Norman contact between the two Celtic countries, but perhaps less between the Anglo-Saxon and Irish Celts.

4   There are very few civilizations which do not have a staple starchy vegetable – those that do not are the few hunter-gatherers who have survived into modern times, the most famous being the Eskimos. The Irish may have had a bread economy until about, say, 500 BC, when the weather changed and they became carnivorous; or they may always have been meat-eaters, who never had a staple starchy diet until the potato appeared.

5   The practice of drawing blood from the living animal may sound grotesque, but it is efficient. A cow gives milk, which may nourish man as well as calf, but 'kine blood' may be drawn from cows, from steers being fattened for meat, from oxen kept for haulage, or from bulls kept for breeding. About 2–3 litres per day may be drawn from an adult animal, about half the volume of milk produced by a cow of the same primitive breed from rough grazing. Other cultures which have drawn blood from the living animal include the Scots, who produced the original haggis from blood and oats; some Lapps, who drew blood from reindeer; and the Masai in Kenya, who still draw blood, in preference to milk, from their cattle.

6   It is difficult today to be accurate about the true origins of the Norsemen. When they burst upon the world they came from Norway, Sweden, Denmark and the Baltic islands. They were men of the soil and of the sea, navigators, shipwrights, fishermen, whalers, yeomen farmers and men who could turn their hand to any trade on earth. Their spiritual side matched the daily needs of their harsh world: pagan, free of rhetorical optimism and false hopes. Their gods were not the encapsulation of the best human virtues, as in Greece or Rome, but mates, fellow players in the game of life, which might end in a great adventure such as death in battle, leading to Valhalla. They were fatalistic, great and good comrades, connoisseurs of war, women, pillage and rape; pirates who admitted of no conscience, no good nor evil, no sin nor virtue. There was no limit to their conduct, no edge to their world.

They were blocked to the south by the strength of the Saxons and Franks, in modern Germany and France, and to the north by ice and permafrost. Any enrichment of the tribe had to be to the east and west, by raids, trading and settlement. All of this was, of course, another name for theft. They were not only thieves but unbelievably cruel. Once a body of raiding Vikings in England complained that they were good men. 'After all,' they said, 'we may have killed all the men and enslaved all the women, but we did not impale all the babies on our swords as we did last year.'

This vigorous race reached and settled most of western Russia, as far as Kiev and Novgorod; they raided, then settled, large parts of England, Scotland, Ireland and Wales, with the exception of the most mountainous regions. They conquered and settled the Faroes, Shetland, the Orkneys and Iceland. They reached North America and only abandoned Greenland when the climate deteriorated. In mainland Europe they conquered and held Normandy and Sicily and sacked the cities of Bordeaux, Seville, Cadiz, Pisa and Valence. The orgy of destruction led in the end to stable, competent, virile feudal government; the city-sackers founded all the towns in Ireland, refounded York, and made Rouen a great city. The conscience-less battle veterans became stern, moralistic Christians, and their administrative skills became, at one remove, the foundation of England's opportunities.

7   Brian Boru (also spelt Bohru, Bohrunna or Boruma) was a native Irish noble, son of Kennedy (Cennetig), who in turn was son of Lorcan, King of Thomond in Cashel. In 977 Brian defeated a conspiracy between Ivar of Limerick, a Scandinavian, and what we would now call collaborators, Donoban and Maelmud. Brian became King of all Munster in 978 and of most of southern Ireland in 984. There was much further fighting as Brian extended his rule by judicious use of the sword. His failure to survive the battle of Clontarf is always seen as the great tragedy of Celtic Ireland, but it is unlikely that the disciplined Norsemen of Ireland, or the Normans (Vikings at one remove), would have tolerated the fey nature of Celtic government without some changes.

8 By means of lordship and overlordship on the Anglo-Norman pattern, Henry II (1133–89) tried to establish feudalism and suzerainty throughout Ireland. He failed, and the Pale was then instituted. It was an area of land which at its maximum comprised eleven shires: Dublin (with Wicklow), Meath (with West Meath), Louth, Carlow, Kilkenny, Wexford, Waterford, Cork, Limerick, Kerry and Tipperary. These counties were established by Henry in 1173–9 and re-established by his son John in 1205–10. They were moderately submissive, depending upon the strength of the Anglo-Norman forces. The feudal oath of obedience, if taken, did not have the same sacred force in Ireland as in the rest of Europe. At its minimum, in 1435–50, the Pale was reduced to an area round Dublin 20 miles deep and 30 miles long. This was the only part of Ireland held by the English at the time. The castles of Wexford, Waterford and Youghal, among many others, were ruined. In the time of Henry VII (1457–1509) the Pale consisted of Dublin, Louth, Kildare and Meath. It was therefore an area whose size was dependent upon the Irish power of the English king.

'Outside' or 'beyond the Pale' became an Anglo-Irish expression denoting unacceptability. Before the Victorians, who used the term indifferently about every conceivable subject, being outside the Pale was a term which would indicate Irishness at best, and semi-criminality at worst. To be outside the Pale for an Englishman before the eighteenth century might well mean literally to risk his life.

9 The tithe was a tax – Jewish, Christian, Arab – assessed as one-tenth of the crops each year. Paid to keep the parish going – not only the priest, but also the fabric of the church – it was abolished in France in 1789, and in England in 1936; it has never existed in the USA. Peter's pence were collected in church and passed to Rome to support the Roman Curia and hierarchy. They are the origin of the modern 'collection' taken at services. The concept of the poor supporting the fabulously rich was much resented. Canonical decrees were the laws laid down from time to time by Rome, and were much disliked by the Irish Church.

10 The Irish, then as now, learnt to hate the worst side of the Anglo-Saxons: philistine, chauvinist, greedy, arrogant, selfish and conceited.

11 The Picts allegedly came from Scythia (around Kiev in the modern Ukraine, in the USSR) in about 4000 BC, going first to Spain and then to Ireland, where they settled in Leinster, with their main concentration in what is now County Meath. Most of them moved to Scotland not long after arriving in Ireland. In Scotland, the Picts tended to settle, or be pushed to the north, the Romans and Saxons occupying the southeast.

The Scots also came from Ireland, but later than the Picts, probably arriving in Scotland just before the Romans. They ended up in the west, and shared the Highlands with the Picts while romanized and anglicized Celts occupied the lowlands. There was also a strong Scots element in that part of Scotland nearest to Ireland and that part of Ireland nearest to Scotland, and traffic across the narrow seas was continuous, if not always peaceful.

12 Elizabeth conducted campaigns in Ireland continuously for twenty years, following a policy of negative imperialism: her aim was to prevent the country falling into the hands of Spain. The history of her Irish wars is complicated and is best explained in the classic work on the subject, Cyril Falls' *Elizabeth's Irish Wars*, Methuen, 1950.

13 The idea of 'plantation' included not only crops and men to work them, but also the master who was granted the right of plantation.

14 The earls who fled at this time were Tyrone, who was harassed and old, accompanied by Tyrconnel and Rory O'Donnell. Cuconnaught Maguire had already left. O'Cahan was judicially ruined. O'Dogherty, chief of Inishowen, was killed, as was O'Hanlon. The Jacobean plantation left not a single Celtic aristocrat in Tyrone, Donegal, Armagh, Cavan, Fermanagh or Derry.

15 Within a century of its introduction into Europe the potato had become settled, native and associated with Ireland. For 150 years before it became a factor in the feeding of other European nations, it had been the staple crop of the Irish peasantry. The white potato was called the Irish potato, and clearly identified with Ireland, in the new American colonies, and American references to 'Irish potatoes' begin in 1635, long before any mutations from Ireland reached America.

16 'Transportation', in the early days to the West Indies, then to mainland North America, and eventually to Australia, was considered a merciful alternative to the gallows as late as the 1860s. In essence, the 'criminal' was 'sold' to a shipmaster who on arrival sold him for seven years' indentured servitude or whatever. After that period the erstwhile criminal was 'free', but denied the right to return to England. Up to a third perished in the ships on the way out, perhaps another third died during their term of servitude, and the remainder, having had all their basic needs supplied for the term of their indenture, found it difficult to become independent free-standing men in the society of their fellow-colonials. The annals of Virginia, for example, abound with stories of this problem. No more than 10–15,000 Catholic Irishmen settled in the mainland colonies during the seventeenth century.

17 Non-Irish potato culture involved planting the setts – whole or cut seed potatoes – then ridging them up to prevent them turning green, and keeping the ridges weed-free throughout the season.

18 'Ascendancy' may be defined as the old Irish aristocracy, become Protestant; the new plantation 'Irishmen' gone native, but remaining rich; or the Anglo-Norman lords, such as Talbot of Malahide, who remained Anglo, but also Irish. To be a member of the Ascendancy it was essential to be landed, rich and relaxed.

19 The status of Ireland was not much different from that of colonial America, the West Indies or India. In mercantile theory the purpose of a colony was to supply the raw materials and the market, and permanently to enrich the metropolitan power, in this case England. See also note 23 to the chapter on Sugar.

20 The continuing persecution of almost every Irish Catholic who could read and write led to a tendency for Ireland to produce a Protestant Anglo–Irish class of intellectuals, among them Swift, Goldsmith, Burke, Wilde and Shaw.

21 There was, of course, a grain trade between the American colonies, which were in surplus, and the West Indian islands, which were in deficit; but the North Atlantic trade was discouraged by every mother country anxious to promote its own agriculture. The first evidence of the future importance of the grain trade was the battle of the Glorious First of June, 1794, in which Admiral Howe of England defeated the French fleet which was convoying grain from America to France. It is typical of the thought of the day, which did not include a grain trade, that Howe destroyed the fleet instead of the convoy, the loss of which might well have produced another revolution in France. Howe was not criticized by the government for this course of action.

22   E.C. Large, *The Advance of the Fungi*, Jonathan Cape, 1940.

23   Free trade is practically impossible without information; education and the training to use it; roads; vehicles; ships; port facilities to handle goods; and good communications in order to learn about prices, quantities, deliveries and the specifications of the goods to be traded. This rules out the less advanced countries.

24   See note 23 to the chapter on Sugar.

25   The Irish Volunteers, perhaps 40,000 in number, were raised in 1778 when France and the American colonies became allies in the War of Independence. The Volunteers soon became a political movement. A political and military convention in 1779 held out for free trade between Britain and Ireland, a demand which was conceded by a British government unwilling to fight both America and Ireland. The Volunteers remained a force in being, a political lever and a threat to English hegemony, until the Act of Union in 1801. It was the identification of the Irish Volunteers with American Independence which first highlighted the coincidence of interest between Ireland and America, long before there was any appreciable number of Irishmen in the colonies.

26   Wolfe Tone (1763–98) founded the Society of United Irishmen in 1791; he visited the United States in 1795 and Revolutionary France in 1796. Tone persuaded the French to invade Ireland on behalf of the patriots, but the French failed to land and Tone himself was captured aboard a French ship in 1798. He was tried and condemned to death, but committed suicide in prison.

27   The Orange Society, set up in secret lodges, was originally founded in 1795 by extreme Protestants to safeguard the Settlement made after William of Orange's success at the battle of the Boyne and as a counterweight to the United Irishmen of Wolfe Tone. It was virtually confined to Ulster.

28   This totals about £15 million going abroad, from an island whose population was 9 million – not much, it could be said, only about £1.66 or less than $2 per head per year. But it was too much, about $1.5 billion in today's money, and equal to the profits from a large bank, or a medium-sized oil company, or half the total annual income of the United Nations. It was, in fact, probably equivalent to all the available rents or profits in Ireland in any one year in the 1840s. The island was being sucked dry.

29   A parallel within living memory can be drawn. In 1943, the hinge year of World War II, a very serious famine in Bengal killed more people than did the whole war east of Suez between 1941 and 1945. Not much attention was paid to the famine, because so many other more 'important', more newsworthy things were happening; more significantly, nothing much could be done to alleviate the shortages. The problem was not availability of food, but the shipping position. In a sense, the Germans were responsible for the famine, since they had sunk so many Allied ships; but it was the British who were blamed by the Indians.

30   All political postulation needs 'scientific proof', which sometimes sways more people than argument does. David Ricardo (1772–1823) and Karl Marx (1818–83) used the same evidence to produce almost diametrically opposed and contrasting scientific 'proofs' of the nature of the value of wages. Both are much quoted by believers, in completely opposite senses.

31   The difference between population increase and food increase can best be explained by a little story of two rabbits. In February of one year there were two rabbits in a grass-filled walled garden of about an acre. Fifty days later there were 14 rabbits; 100 days later, 97 rabbits; 150 days later, 682 rabbits; and 200 days later, 3300 plus. It was now September, and the grass was not growing all that well.

Long before, the rabbits had exceeded the capacity of the acre of grass to support grass-eating animals. Though the grass had responded, since it grows better when cropped short, the productivity of the acre had only doubled. Humans are not rabbits, but they can always breed faster than the supply of food can be increased.

32   But famines still occur. India was the great postwar problem, and Malthus played his part as late as 1970. 'You'll only get wheat,' said the Indian Famine Relief Director to Bengali villagers in that year, 'if an interuterine coil is installed for every sack of wheat I give you.' This was a direct, mathematical application of Malthusian principles.

33   This volte-face split Peel's party into two: Peelites, such as Gladstone, followed their leader and were known as 'liberal conservatives', while Disraeli seized his politcal opportunity and brought stinging rebukes upon his former leader's head.

34   In the eighteenth century no one pretended that moral and political questions were inevitably connected. Some were, some weren't; others were matters of convenience or amenity. The new Liberals of 1832 and after, then the socialists and then the Communists, all pretended that their case was a moral one while that of the Opposition was selfish, obscurantist and blatantly bad. This assumption had a very powerful effect in Victorian England which lasted in Parliament until 1914. Efforts to revive the connection in more recent times have been ridiculed.

35   A day-worker is paid for the time he spends on the job. A piece-worker is paid for the particular task he has done. In the early, pre-industrial eighteenth century the English worker could not compete with his foreign counterpart in any aspect of cotton manufacture, so tariffs were introduced to protect English wage levels.

36   It is worth noting that moving earth, or stone, or minerals is one of the few tasks which is absolutely cheaper than it was fifty years ago – such are the efficiencies of modern machinery. Thus an open-cast or drift mine must by its very nature be much cheaper to work than a deep mine. This is why no European deep-mined coal can compete on the world market, now or in the future. It is an example of straight physical facts, rather than the voter at the ballot box, dictating political and economic results.

37   A steamship used to be divided into three classes of accommodation. In the fo'c'sle (forecastle), in the bows, subject to heavy pitching, were the crew. Amidships were the officers and the first-class passengers, at the fulcrum point for both pitching and rolling – the most comfortable place in the ship in the days before stabilizers. In the stern, the rudder and machinery were down below, in the so-called steerage; this is where the emigrants were huddled. They paid as little as one-tenth the price of a first-class ticket in an Atlantic liner. Most English ships crossing to America called at Cork to pick up Irish emigrants; alternatively the Irish crossed to Liverpool to start their voyage, as did many emigrants from Europe. This was a regular trade, promoted not only by shipping lines from, say, Hamburg to Hull, but also by the railway companies involved.

Bibliography

Certain works which have been cited in the Notes are not repeated here. *Bibliotheca Americana*, also not listed, purports to contain every work about the Americas published between 1496 and 1783; copies can be found in the Library of Congress and in the British Library. Finally, a word on behalf of libraries and librarians: most libraries contain a useful subject index, and librarians are much more than random access memories. No librarian, given the chance, has ever failed to help the author.

Altschul, Siri von Reis, 'Exploring the Herbarium', *Scientific American*, May 1977
Alvord, C.W. *The Mississippi Valley in British Politics*, Cleveland, 1917
Anderson, A., *An Historical and Chronological Deduction of the Origin of Commerce*, rev. edn, 4 vols, London, 1787–9
Anderson, Edgar, *Plants, Man and Life*, London, 1954
Anderson, S., *The Sailing Ship*, New York, 1947
Andrews, C.M., *The Colonial Period of American History*, 4 vols, New Haven, 1938
Anstey, Roger, *The Atlantic Slave Trade and British Abolition 1760–1810*, London, 1975
Arnold, Sir Thomas, and Alfred Guillaume, *The Legacy of Islam*, London, 1931
Atterbury, P., ed., *The History of Porcelain*, London, 1982
Barbour, Violet, *Capitalism in Amsterdam in the 17th Century*, Ann Arbor, 1963
Bastin, J., *The British in West Sumatra 1685–1825*, Kuala Lumpur, 1965
Berlin, Isaiah, *Four Essays on Liberty*, London, 1969
Berlin, Isaiah, *Historical Inevitability*, London, 1953
Biddulph, J., *The Pirates of Malabar*, London, 1907
Bloch, Marc, *Feudal Society*, London, 1961
Boswell, James, ed. Augustine Birrell, *The Life of Samuel Johnson*, 6 vols, London, 1904
Bougainville, L.A. de, *Voyage Autour du Monde*, Paris, 1771
Boxer, C.R., *The Dutch Seaborne Empire*, London, 1965
Brougham, Henry Peter, *An Inquiry into the Colonial Policy of the European Powers*, Edinburgh, 1803; New York, 1969
Bruce, J., *Annals of the East India Company*, London, 1810
Butzer, Karl W., *Early Hydraulic Civilisation in Egypt*, Chicago, 1976
Cambridge Ecomomic History
Cambridge History of England

Campbell, John, *The Spanish Empire in America, by an English Merchant*, London, 1747

Campbell, W., *Formosa under the Dutch*, London, 1903; New York, 1970

Chambers, J.D., and G.E. Murgay, *The Agricultural Revolution*, London, 1966

Champion, Richard, *Considerations on the Present Situation of Great Britain and the United States*, London, 1784

Chang, Kwang-Chih, *The Archaeology of Ancient China*, New Haven, 1978

Childe, V. Gordon, *The Dawn of European Civilisation*, 6th edn, revised, London, 1973

Clapham, Sir John, *An Economic History of Modern Britain*, 3 vols, Cambridge, 1926–38

Clapham, Sir John, *The Ecomonic Development of France and Germany 1815–1914*, 4th edn, Cambridge, 1966

Clark, Colin, *Population Growth and Land Use*, London, 1967

Cook, James, ed. J.C. Beaglehole, *Journals*, 3 vols, Cambridge, 1966

Crane, Eva, *The Archaeology of Beekeeping*, London, 1983

Curtin, Philip D., *The Atlantic Slave Trade*, Madison, 1969

Dalrymple, A., *An Historical Collection of the Several Voyages and Discoveries in the South Pacific Ocean*, 2 vols, London, 1770–1; New York, 1967

Dampier, William, *A New Voyage Round the World*, 3 vols, London, 1697; New York, 1968

Davis, David Brion, *The Problem of Slavery in Western Culture*, Ithaca, 1966

Davis, Ralph, *The Rise of the English Shipping Industry in the Seventeenth and Eighteenth Centuries*, London, 1962; New York, 1963

Deane, Phyllis, and W.A. Cole, *British Economic Growth 1688–1959*, Cambridge, 1967

Deerr, N., *The History of Sugar*, Oxford, 1950

Dodge, Ernest S., *New England and the South Seas*, Cambridge, Mass., 1965

Drummond, J.C., *The Englishman's Food*, London, 1957

Ernle, Lord (Rowland Prothero), *English Farming, Past and Present*, 4th edn, London, 1927

Fisher, H.A.L., *A History of Europe*, London, 1936

Franklin, Benjamin, ed. A.M. Smyth, *Writings*, 10 vols, New York, 1907

Freyre, Gilberto, *The Masters and the Slaves*, New York, 1970

Galbraith, V.H., *Domesday Book – Its Place in Administrative History*, Oxford, 1974

Gibbon, Edward, *The History of the Decline and Fall of the Roman Empire*, London, 1969

Gipson, L.H., *The British Empire before the American Revolution*, vol. X, *The Triumphant Empire; Thunder-clouds Gather in the West, 1763–1768*, New York, 1961

Gray, L.C., *The History of Agriculture in the Southern States to 1860*, Washington, DC, 1933

Greenberg, M., *British Trade and the Opening of China*, Cambridge, 1951

Harler, *The Culture and Marketing of Tea*, Oxford, 1955

Harlow, V.T., *The Founding of the Second British Empire*, 2 vols, London and New York, 1952 and 1964

Harrison, Gordon, *Mosquitoes, Malaria and War. A History of the Hostilities since 1880*, New York, 1978

Harvard Bibliography of American History

Hawkesworth, J., *Voyages in the Southern Hemishpere*, 3 vols, London, 1773

Historical Statistics of the US, Washington, DC, 1960

Hobsbawm, E.J., *Industry and Empire*, London and New York, 1968

Hourani, G.F., *Arab Seafaring in the Indian Ocean in Ancient and Early Medieval Times*, Princeton, 1951

Howard, Michael, 'Power at Sea', *Adelphi Papers*, no. 124, London, 1976

Howard, Michael, *War in European History*, Oxford, 1975

Huntington, Ellsworth, *Civilisation and Climate*, New Haven, 1924

Jenks, Leland, *Our Cuban Colony: A Study in Sugar*, New York, 1928

Klerck, E.S. de, *History of the Netherlands East Indies*, 2 vols, Rotterdam, 1938

Labaree, L.W., *Royal Government in America: A Study of the British Colonial System before 1783*, New Haven, 1930

Ligon, R., *A True and Exact History of the Island of Barbados*, London, 1657; New York, 1970

Loomis, Robert S., 'Agricultural Systems', *Scientific American*, September 1976

Lynch, John, *The Spanish American Revolutions*, London, 1973

Marwick, Arthur, *Britain in the Century of Total War*, London, 1968

Medawar, P.B., *The Hope of Progress*, London, 1974

Milburn, W., *Oriental Commerce; containing a geographical description of the principal places in the East Indies . . . with their Produce, Manufactures and Trade . . .*, 2 vols, London, 1813

Mitchell, B.R., *European Historical Statistics 1750–1970*, London, 1975

Mitchell, B.R., with Phyllis Deane, *Abstract of British Historical Statistics*, Cambridge, 1962

Morison, Samuel Eliot, *Christopher Columbus, Mariner*, London, 1956

Morse, H.B., *The Chronicles of the East India Company Trading in China, 1635–1834*, Oxford, 1926; New York, 1965

Namier, Sir Lewis, *Crossroads of Power*, London, 1962

Nicolson, Harold, *The Congress of Vienna*, London, 1946

Norwood, Richard, *The Seaman's practice, containing a fundamental Problem in Navigation, experimentally verified, viz: touching the Compass of the Earth and Sea, and the Quantity of a Degree in our English Measure, also to keep a reckoning at Sea for all Sailing, etc. etc.*, London, 1637

Oman, Sir C., *A History of the Art of War in the Middle Ages*, London, 1924

Origo, Iris, *The Merchant of Prato*, London, 1957

Orwin, C.S., *The Open Fields*, Oxford, 1967

Oxford History of Technology

Parry, J.H., *The Age of Reconnaissance*, London, 1963

Parry, J.H., *The Spanish Seaborne Empire*, London, 1966

Parry, J.H., *Trade and Dominion*, London, 1971

Philips, C.H., *The East India Company 1784–1834*, Manchester, 1940

Plato, *The Republic*

Plumb, J.H., *Man versus Society in Eighteenth Century England*, London, 1969

Pole, J.R., *Political Representation in England and the Origins of the American Republic*, London, 1966

Price, A. Grenfell, *The Western Invasion of the Pacific and Its Continents*, Oxford, 1963

Read, Herbert, *Anarchy and Order: Essays in Politics*, London, 1954

Rostow, Walt, ed., *The Economics of 'Take Off' into Self-sustained Growth*, Washington, DC, 1963

Rousseau, Jean-Jacques, *A Discourse on the Origin of Inequality*, London, 1952

Russell, E. John, *The World of the Soil*, London, 1961

Sadler, D.H., intro., *Man Is not Lost: a record of 200 years of astronomical navigation with the Nautical Almanac, 1767–1967*, London, 1968

Salaman, R., *The History and Social Influence of the Potato*, Cambridge, 1949

Sauvy, Alfred, *General Theory of Population*, London, 1969

Sawyer, P.H., *The Age of Vikings*, London, 1962

Steel, David, *Elements and Practice of Rigging and Seamanship*, 2 vols, London, 1794

Steel, David, *Elements and Practice of Naval Architecture*, 2 vols, London, 1805

Tacitus, *Histories*, trans. Kenneth Wellesley, London, 1964

Tawney, R.H., *Religion and the Rise of Capitalism*, London, 1969

Tench-Cox, *A View of the United States of America*, Philadelphia, 1794

Thomas, Hugh, *Cuba or the Pursuit of Freedom*, London, 1971

Thomas, Hugh, *An Unfinished History of the World*, London, 1979

Thompson, E.A., *The Early Germans*, Oxford, 1965

Thompson, E.P., *The Making of the English Working Class*, London, 1965

Trevelyan, G.M., *English Social History*, London, 1942

Tucker, Josiah, *The True Interest of Britain, set forth in regard to the Colonies: and the only means of living in peace and harmony with them*, London, 1774

Wadia, R.A., *The Bombay Dockyard and the Wadia Master-builders*, Bombay, 1957

Walpole, Horace, *Memoirs of the Reign of King George the Third*, London, 1845

Weber, Max, *The Protestant Ethic and the Spirit of Capitalism*, London, 1930

Washington, George, ed. John C. Fitzpatrick, *Writings*, 39 vols, Washington, DC, 1931–44

Williams, E.T., *A Short History of China*, New York, 1928

Williams, Glyndwr, *The British Search for the Northwest Passage in the Eighteenth Century*, London, 1962; New York, 1967

Index